Physics of Spin in Solids: Materials, Methods and Applications

NATO Science Series

A Series presenting the results of scientific meetings supported under the NATO Science Programme.

The Series is published by IOS Press, Amsterdam, and Kluwer Academic Publishers in conjunction with the NATO Scientific Affairs Division

Sub-Series

I. Life and Behavioural Sciences	IOS Press
II. Mathematics, Physics and Chemistry	Kluwer Academic Publishers
III. Computer and Systems Science	IOS Press
IV. Earth and Environmental Sciences	Kluwer Academic Publishers
V. Science and Technology Policy	IOS Press

The NATO Science Series continues the series of books published formerly as the NATO ASI Series.

The NATO Science Programme offers support for collaboration in civil science between scientists of countries of the Euro-Atlantic Partnership Council. The types of scientific meeting generally supported are "Advanced Study Institutes" and "Advanced Research Workshops", although other types of meeting are supported from time to time. The NATO Science Series collects together the results of these meetings. The meetings are co-organized bij scientists from NATO countries and scientists from NATO's Partner countries – countries of the CIS and Central and Eastern Europe.

Advanced Study Institutes are high-level tutorial courses offering in-depth study of latest advances in a field.
Advanced Research Workshops are expert meetings aimed at critical assessment of a field, and identification of directions for future action.

As a consequence of the restructuring of the NATO Science Programme in 1999, the NATO Science Series has been re-organised and there are currently Five Sub-series as noted above. Please consult the following web sites for information on previous volumes published in the Series, as well as details of earlier Sub-series.

http://www.nato.int/science
http://www.wkap.nl
http://www.iospress.nl
http://www.wtv-books.de/nato-pco.htm

Series II: Mathematics, Physics and Chemistry – Vol. 156

Physics of Spin in Solids: Materials, Methods and Applications

edited by

Samed Halilov
Naval Research Laboratory,
Washington, DC
and University of Pennsylvania,
Philadelphia, U.S.A.

Kluwer Academic Publishers

Dordrecht / Boston / London

Published in cooperation with NATO Scientific Affairs Division

Proceedings of the NATO Advanced Research Workshop on
Physics of Spin in Solids: Materials, Methods and Applications
Baku, Azerbaijan
15–19 October 2003

A C.I.P. Catalogue record for this book is available from the Library of Congress.

ISBN 1-4020-2225-5
ISBN 1-4020-2708-7 (e-book)

Published by Kluwer Academic Publishers,
P.O. Box 17, 3300 AA Dordrecht, The Netherlands.

Sold and distributed in North, Central and South America
by Kluwer Academic Publishers,
101 Philip Drive, Norwell, MA 02061, U.S.A.

In all other countries, sold and distributed
by Kluwer Academic Publishers,
P.O. Box 322, 3300 AH Dordrecht, The Netherlands.

Printed on acid-free paper

Printed in the Netherlands.

Contents

Contributing Authors

B. L. Györffy, H. H. Wills Physics Laboratory, University of Bristol, Tyndall Ave., Bristol BS8 1TL, UK
Centre for Computational Materials Science, TU Wien, Gertreidemarkt 9/134, A-1060 Wien,Austria

M. Krawiec, Institute of Physics and Nanotechnology Center, Maria Curie-Skłodowska University, Pl. Marii Curie-Skłodowskiej 1, 20-031 Lublin, Poland

J. F. Annett, H. H. Wills Physics Laboratory, University of Bristol, Tyndall Ave., Bristol BS8 1TL, UK

Hiroyoshi Momida, National Institute for Materials Science, 1-2-1 Sengen, Tsukuba 305-0047, Japan

Tamio Oguchi, ADSM, Hiroshima University
1-3-1 Kagamiyama, Higashihiroshima 739-8530
Japan

A. M. Gabovich, Institute of Physics, prospekt Nauki 46, 03028 Kiev-28, Ukraine

A. I. Voitenko, Institute of Physics, prospekt Nauki 46, 03028 Kiev-28, Ukraine

Mai Suan Li, Institute of Physics, Al. Lotnikow 32/46, PL-02-668 Warsaw, Poland

H. Szymczak, Institute of Physics, Al. Lotnikow 32/46, PL-02-668 Warsaw, Poland

M. Pekala, Department of Chemistry, University of Warsaw, Al. Zwirki i Wigury 101, PL-02-089 Warsaw, Poland

Silke Biermann, Centre de Physique Theorique, Ecole Polytechnique, 91128 Palaiseau, France

Ferdi Aryasetiawan, Research Institute for Computational Sciences, AIST,
1-1-1 Umezono, Tsukuba Central 2, Ibaraki 305-8568, Japan

Antoine Georges, Centre de Physique Theorique,
Ecole Polytechnique, 91128 Palaiseau, France

S.L. Molodtsov, Institut fr Festkrperphysik, Technische Universitt Dresden,
D-01062 Dresden, Germany

M.D. Johannes, Center for Computational Material Science
Naval Research Laboratory
Washington, D.C. 20375

Siyavush Azakov, Institute of Physics, Azerbaijan Academy of Sciences, Baku-370143, Azerbaijan

Nicholas Kioussis, Department of Physics
California State University Northridge, California 91330-8268

Yan Luo, Department of Physics
California State University Northridge, California 91330-8268

Claudio Verdozzi, Department of Physics
California State University Northridge, California 91330-8268

F. M. Grosche, Department of Physics
Royal Holloway, University of London, Egham, UK

D. Moroni-Klementowicz, Department of Physics
Royal Holloway, University of London, Egham, UK

R. Burrell, Department of Physics
Royal Holloway, University of London, Egham, UK

D. Fort, School of Engineering, Metallurgy and Materials,
University of Birmingham, Birmingham, UK

J. Awaka, Department of Materials Science and Engineering,
Muroran Institute of Technology, 27-1 Mizumoto-cho, Muroran, Hokkaido 050-8585, Japan

S. Nagata, Department of Materials Science and Engineering,
Muroran Institute of Technology, 27-1 Mizumoto-cho, Muroran, Hokkaido 050-8585, Japan

H. Q. Yuan, Max-Planck-Institute for Chemical Physics of Solids,
Nöthnitzer Str. 40, D-01187 Dresden, Germany

M. Deppe, Max-Planck-Institute for Chemical Physics of Solids,
Nöthnitzer Str. 40, D-01187 Dresden, Germany

C. Geibel, Max-Planck-Institute for Chemical Physics of Solids,
Nöthnitzer Str. 40, D-01187 Dresden, Germany

G. Sparn, Max-Planck-Institute for Chemical Physics of Solids,
Nöthnitzer Str. 40, D-01187 Dresden, Germany

F. Steglich, Max-Planck-Institute for Chemical Physics of Solids,
Nöthnitzer Str. 40, D-01187 Dresden, Germany

Lars Nordström, Department of Physics,
Uppsala University, Box 530, S-751 21 Uppsala, Sweden

Till Burkert, Department of Physics,
Uppsala University, Box 530, S-751 21 Uppsala, Sweden

Massiliano Colarieti-Tosti, Department of Physics,
Uppsala University, Box 530, S-751 21 Uppsala, Sweden

Olle Eriksson, Department of Physics,
Uppsala University, Box 530, S-751 21 Uppsala, Sweden

I.I. Mazin,Center for Computational Materials Science
Naval Research Laboratory
Washington, DC 20375, USA

D.J. Singh,Center for Computational Materials Science
Naval Research Laboratory
Washington, DC 20375, USA

A. Aguayo,Center for Computational Materials Science
Naval Research Laboratory
Washington, DC 20375, USA

I. N. Askerzade, Institute of Physics, Azerbaijan National Academy of Sciences, Baku-370143, Azerbaijan

A. Perlov,University of Munich, Butenandstrasse 5-13, D-81377, Munich, Germany

S. Chadov,University of Munich, Butenandstrasse 5-13, D-81377, Munich, Germany

H. Ebert,University of Munich, Butenandstrasse 5-13, D-81377, Munich, Germany

L. Chioncel,University of Nijmegen, NL-6526 ED Nijmegen, The Netherlands

A.Lichtenstein,University of Nijmegen, NL-6526 ED Nijmegen, The Netherlands

M. Katsnelson,Uppsala University, P.O.Box 530, S-751 21 Uppsala, Sweden

K. I. Kugel, Institute for Theoretical and Applied Electrodynamics, Russian Academy of Sciences
Izhorskaya str. 13/19, Moscow 125412, Russia

A. L. Rakhmanov, Institute for Theoretical and Applied Electrodynamics, Russian Academy of Sciences
Izhorskaya str. 13/19, Moscow 125412, Russia

A. O. Sboychakov, Institute for Theoretical and Applied Electrodynamics, Russian Academy of Sciences Izhorskaya str. 13/19, Moscow 125412, Russia

M. Yu. Kagan, Kapitza Institute for Physical Problems, Russian Academy of Sciences Kosygina str. 2, Moscow, 119334 Russia

I. V. Brodsky, Kapitza Institute for Physical Problems, Russian Academy of Sciences Kosygina str. 2, Moscow, 119334 Russia

A. V. Klaptsov, Kapitza Institute for Physical Problems, Russian Academy of Sciences Kosygina str. 2, Moscow, 119334 Russia

E. M. Kerimova, Institute of Physics, National Academy of Sciences of Azerbaijan, Baku-370143, Azerbaijan

S. N. Mustafaeva, Institute of Physics, National Academy of Sciences of Azerbaijan, Baku-370143, Azerbaijan

A. I. Jabbarly, Institute of Physics, National Academy of Sciences of Azerbaijan, Baku-370143, Azerbaijan

G. Sultanov, Institute of Physics, National Academy of Sciences of Azerbaijan, Baku-370143, Azerbaijan

A.I.Gasanov, Institute of Physics, National Academy of Sciences of Azerbaijan, Baku-370143, Azerbaijan

R. N. Kerimov, Institute of Physics, National Academy of Sciences of Azerbaijan, Baku-370143, Azerbaijan

R.T. Tagiyeva, Institute of Physics Azerbaijan National Academy of Sciences, Baku-370143, Azerbaijan

S. V. Halilov, Center for Computational Materials Science, Naval Research Laboratory, Washington, DC 20375, and Department of Materials Science and Engineering, University of Pennsylvania, Philadelphia, PA 19104

Preface

In 1925, H. Ulenbeck and S. Goudsmith suggested the idea of spin as a kind of intrinsic moment in electrons, thus establishing a fundamental building block in the physics of magnetic phenomena in solids. Currently, research in the area of magnetism is largely motivated by the needs of technology. However, research efforts outside the scope of technological needs are useful for the advancement of the physics of spin itself. It is well known that spin-dependent/related phenomena such as magneto-resistance, superconductivity, magneto-optics and others have been under intense theoretical and experimental investigation resulting in the creation of new devices in the area of digital recording and spin-dependent transport-based technologies. On the other hand, further development of experimental research techniques enhanced the accuracy of measurements such as, for example, scanning tunneling microscopy, exchange force microscopy, angle- and spin-resolved photoemission, scattering of neutrons and electrons on magnetic type of excitations, challenging physicists to improve the interpretations of results on a computational rather than qualitative basis. Some examples are new quantitative understandings of phenomena such as the interplay of different types of low-lying excitations involving charge, spin, and lattice degrees of freedom in manganites and conventional superconductors, the unique role of quantum spin fluctuations in thermodynamics and transport properties of unconventional superconductors and weak ferromagnets, the influence of electron correlations on order/disorder of magnetic nanoclusters or quantum dots, and co-existance of non-uniform magnetic ordering and superconductivity.

This book reviews a rather limited selection of recent progress made in the vast field of magnetism. The list of chapters comprise such issues as: new many-body theoretical developments and applications designed to more accurately describe the ground state and excitations of strongly correlated systems on arbitrary energy scales, spin-dependent tunnel currents and proximity effects at the interface of magnetic materials and superconductors, application of first-principle calculations to explore spin-dependent material properties close to a quantum critical point such as spectrum of magnetic fluctuations, magneto-crystalline anisotropy, non-uniform magnetism, search for quantum phase transitions in narrow-band metals.

SAMED HALILOV

Washington, D.C.
February, 2004

Acknowledgments

All participants of the workshop are grateful to the NATO Public Diplomacy Division in Brussels for guidance and financial support. The help of Dr. F. Pedrazzini, NATO Scientific Programme Director, is highly appreciated. Our special thanks also to Office of Naval Research, International Field Office in London, which kindly agreed to co-sponsor the meeting. We are thankful to Dr. H. Pashaev, Ambassador of the Embassy of the Republic of Azerbaijan in Washington, D.C. and the Institute of Physics in Baku, for their hospitality and efficient organization of the meeting, which stimulated many useful discussions between various research groups from different countries.

FULDE-FERRELL-LARKIN-OVCHINNIKOV-LIKE STATE IN FERROMAGNET - SUPERCONDUCTOR PROXIMITY SYSTEM

B. L. Györffy

H. H. Wills Physics Laboratory, University of Bristol, Tyndall Ave., Bristol BS8 1TL, UK
Centre for Computational Materials Science, TU Wien, Gertreidemarkt 9/134, A-1060 Wien, Austria

M. Krawiec

Institute of Physics and Nanotechnology Center, Maria Curie-Skłodowska University, Pl. Marii Curie-Skłodowskiej 1, 20-031 Lublin, Poland

J. F. Annett

H. H. Wills Physics Laboratory, University of Bristol, Tyndall Ave., Bristol BS8 1TL, UK

Abstract We discuss some properties of the ferromagnet - superconductor proximity system. In particular, the emphasis is put on the physics of the Fulde-Ferrell-Larkin-Ovchinnikov ($FFLO$) like state. In addition to Andreev reflections it features a number of unusual thermodynamic and transport properties, like: oscillatory behavior of the pairing amplitude, density of states and superconducting transition temperature as a function of the ferromagnet thickness. Surprisingly, under certain conditions spontaneous spin polarized current is generated in the ground state of such a system. We provide some informations regarding experimental observations of this exotic state.

Keywords: Proximity effect, Fulde-Ferrell-Larkin-Ovchinnikov state

S. Halilov (ed.), Physics of Spin in Solids: Materials, Methods and Applications, 1–16.

1.1 Introduction

When a normal non-magnetic metal is connected to a superconductor it acquires superconducting properties, like non-zero pairing amplitude. This effect, known as the proximity effect [1], has extensively been studied for almost half a century. It is rather well understood by now in terms of Andreev reflections [2], according to which an impinging electron (with energy less than superconducting gap) on the normal metal (NM) / superconductor (SC) interface is reflected back as a hole and the Cooper pair is created in superconductor. From the point of view of Andreev reflections the proximity effect can be regarded as a non-zero density of the Andreev correlated electron - hole pairs on the normal metal side of the interface.

When a normal metal is replaced by a ferromagnet (FM), another energy scale enters problem, namely the exchange splitting which is related to the spin polarization of the electrons. Such FM/SC hybrid structures are important from the scientific point of view, as they allow the study of the interplay between ferromagnetism and superconductivity [3] as well as of device applications in such areas of technology as magnetoelectronics [4] or quantum computing [5].

It is widely accepted that ferromagnetism and superconductivity are two antagonistic phenomena, so one could expect that the proximity effect in FM/SC system should be suppressed. Indeed, the one can argue that in ferromagnet there are different numbers of spin-up (majority) $n_\uparrow$ and spin-down (minority) $n_\downarrow$ conduction channels, and due to the fact that incident and reflected particles occupy different spin bands, only a fraction $n_\downarrow/n_\uparrow$ of majority particles can be Andreev reflected [6].

On the other hand if an exchange field acts on the Cooper pairs, one would expect that either it is too weak to break the pair, or it suppresses completely superconductivity. However when a Cooper pair is subjected to the exchange field, it acquires a finite momentum and for certain values of the exchange splitting a new superconducting state is realized, known as *Fulde - Ferrell - Larkin - Ovchinnikov* ($FFLO$) state [7, 8]. Interestingly such state features a spatially dependent order parameter corresponding to the non-zero center of mass motion of the Cooper pairs. This state features in non-zero spin polarization, almost normal tunneling characteristics and almost normal Sommerfeld specific heat ratio, anisotropic electrodynamic properties. Unfortunately the bulk state is very sensitive to the impurities and shape of the Fermi surface. Another novel feature of this state is a current flowing in the ground state. The unpaired electrons tend to congregate at one portion of the Fermi surface so a quasiparticle current is produced. In order to satisfy the Bloch theorem: no current in the ground state, a supercurrent, generated by the nonzero value of the pairing momentum, flows in opposite direction, and the total current is zero.

Similar oscillations of the pairing amplitude have been predicted [9]-[12] in ferromagnet/superconductor proximity systems. It turns out that these oscillations are responsible for the oscillatory behavior of the SC critical temperature T_c, first experimentally observed by Wong *et al.* [13], and the density of states [14] as the thickness of the FM slab is varied. In fact, the oscillations of the T_c in FM/SC multilayers can be also explained in terms of the effective π-junction behavior [10]. It was shown that at specific FM thickness the Josephson coupling between two SC layers can lead to a junction with an intrinsic phase (of the order parameter) difference $\delta\varphi = \pi$, which exhibits a higher T_c than the ordinary one ($\delta\varphi = 0$). The π-junction effect has been originally proposed by Bulaevskii *et al.* [15] to arise in the tunnel barriers containing magnetic impurities. It was also suggested that the π-junction can be realized in high-T_c superconducting weak links [16], where the SC order parameter changes its sign under $\pi/2$ rotation. This has tremendous consequences as it leads to many important effects [17, 18], like: the zero energy Andreev states, zero-bias conductance peaks, large Josephson current, time reversal symmetry breaking, paramagnetic Meissner effect and spontaneously generated currents.

From the point of view of the present paper the important issue is the formation of the Andreev bound states in FM/SC proximity system. The Andreev states arise due to the fact that the quasiparticles of the ferromagnet participating in the Andreev reflections move along closed orbits. Such states have been first studied by de Gennes and Saint-James [19] in the insulator/normal metal/superconductor ($I/NM/SC$) trilayer. The energies of these states are always smaller than the SC gap Δ and symmetrically positioned around the Fermi level. They strongly depend on the geometry of the system as well as on the properties of the interfaces. In high-T_c (d-wave) superconductors, these states can be shifted to zero energy, due to the specific form of the symmetry of the order parameter [20], thus indicating π-junction behavior in the system. Naturally, such Andreev states can also arise in the $I/FM/SC$ heterostructures. Moreover, it is possible to shift the energies of these states by changing the exchange splitting, as was first demonstrated by Kuplevakhskii & Fal'ko [21]. In turn, by properly adjusting the exchange splitting the position of the Andreev bound states can be moved to the Fermi energy. The system under such circumstances behaves like that being in the π-junction phase as the spontaneous current is generated [22].

Some of our results have already been published [22]-[24]. Here we wish to present a more detailed study of the FM/SC proximity system in terms of $FFLO$ physics. In some situations the ground state of FM/SC structures has properties of both the $FFLO$ and the π-junction, leading to various interesting and unexpected phenomena.

The paper is organized as follows: In Sec. 1.2 the simple model which allows for self-consistent description of the FM/SC heterostructure is introduced. In Sec. 1.3 the nature of the Andreev bound states in the ferromagnet is discussed. The spontaneously generated current and corresponding magnetic field in the ground state are studied in the Sec. 1.4. In Sec. 1.5 show some transport properties of the system, in the Sec. 1.6 we compare our system to usual $FFLO$ state, and finally, we conclude in Sec. 1.6.

1.2 Model and theory

To study the properties of FM/SC system we have adopted the $2D$ Hubbard model featuring the exchange splitting in the ferromagnet and an electron - electron attraction in superconductor. The Hamiltonian is:

$$H = \sum_{ij\sigma} \left[t_{ij} + \left(\frac{1}{2} E_{ex}\sigma - \mu \right) \delta_{ij} \right] c^{+}_{i\sigma} c_{j\sigma} + \frac{1}{2} \sum_{i\sigma} U_i n_{i\sigma} n_{i-\sigma} \tag{1}$$

where in the presence of a vector potential $\vec{A}(\vec{r})$, the hopping integral is given by $t_{ij} = -te^{-ie\int_{\vec{r}_i}^{\vec{r}_j} \vec{A}(\vec{r})\cdot d\vec{r}}$ for nearest neighbor lattice sites, whose positions are $\vec{r}_i$ and $\vec{r}_j$, and zero otherwise. The exchange splitting E_{ex} is only non-zero on the FM side, unlike as U_i (electron - electron attraction) being non-zero only in SC. μ is the chemical potential, $c^{+}_{i\sigma}$, $(c_{i\sigma})$ are the usual electron creation (annihilation) operators and $\hat{n}_{i\sigma} = c^{+}_{i\sigma} c_{i\sigma}$.

In the following we shall work within Spin - Polarized - Hartree - Fock - Gorkov ($SPHFG$) approximation [22] assuming periodicity in the direction parallel to the interface while working in a real space in the direction perpendicular. Labeling the layers by integer n and m at each k_y point of the Brillouin zone we shall solve the following $SPHFG$ equation:

$$\sum_{m',\gamma,k_y} H^{\alpha\gamma}_{nm'}(\omega, k_y) G^{\gamma\beta}_{m'm}(\omega, k_y) = \delta_{nm}\delta_{\alpha\beta} \tag{2}$$

where the only non-zero elements are: H^{11}_{nm} and $H^{22}_{nm} = (\omega - \frac{1}{2}\sigma E_{ex} \pm \mu \pm tcos(k_y \mp eA(n)))\delta_{nm} \pm t\delta_{n,n+1}$ for the upper and lower sign respectively, $H^{33}_{nm} = H^{11}_{nm}$ and $H^{44}_{nm} = H^{22}_{nm}$ with σ replaced by $-\sigma$ and $H^{12}_{nm} = H^{21}_{nm} = -H^{34}_{nm} = -H^{43}_{nm} = \Delta_n \delta_{nm}$ and $G^{\alpha\beta}_{nm}$ is corresponding Green's function (GF).

As usual, the self-consistency is assured by the relations determining the FM (m_n) and SC (Δ_n) order parameters, current ($J_{y\uparrow(\downarrow)}(n)$) and the vector potential ($A_y(n)$) respectively:

$$m_n = n_{n\uparrow} - n_{n\downarrow} = \frac{2}{\beta} \sum_{ky} \sum_{\nu=0}^{2N-1}$$
$$\mathrm{Re}\left\{ (G_{nn}^{11}(\omega_\nu, k_y) - G_{nn}^{33}(\omega_\nu, k_y)) e^{(2\nu+1)\pi i/2N} \right\} \quad (3)$$

$$\Delta_n = U_n \sum_{k_y} \langle c_{n\downarrow}(k_y) c_{n\uparrow}(k_y) \rangle$$
$$= \frac{2U_n}{\beta} \sum_{ky} \sum_{\nu=0}^{2N-1} \mathrm{Re}\left\{ G_{nn}^{12}(\omega_\nu, k_y) e^{(2\nu+1)\pi i/2N} \right\} \quad (4)$$

$$J_{y\uparrow(\downarrow)}(n) = \frac{4et}{\beta}$$
$$\times \sum_{k_y} sin(k_y - eA_y(n)) \sum_{\nu=0}^{2N-1} \mathrm{Re}\left\{ G_{nn}^{11(33)}(\omega_\nu, k_y) e^{(2\nu+1)\pi i/2N} \right\} \quad (5)$$

$$A_y(n+1) - 2A_y(n) + A_y(n-1) = -4\pi J_y(n) \quad (6)$$

The details of the calculations can be found in [23].

1.3 Andreev bound states

Before we discuss results of fully self-consistent calculations we would like to turn the attention to origin of Andreev bound states and take a look at physics of them from the point of view of semiclassical approach.

From quasiclassical considerations, each bound state corresponds to quasiparticle moving along a family of closed trajectories [25]. The energy of such bound state is determined by the Bohr-Sommerfeld quantization rules, according to which the total phase accumulated during one cycle has to be equal to multiples of 2π. Interestingly, the bound states also emerge in the normal metal/superconductor (NM/SC) structures [19] due to the Andreev reflections [2], according to which an incident electron is reflected back as a hole at the interface, and a Cooper pair is created in SC. Such states are built up from a combination of electron and hole wave functions. The example of the closed quasiparticle trajectory, producing the bound state, in an insulator/(normal metal)/superconductor $I/NM/SC$, is shown in the Fig. 1. It consists of an electron e segment, which includes a ordinary reflection at the I/NM interface, and hole h one, retracing backwards the electron

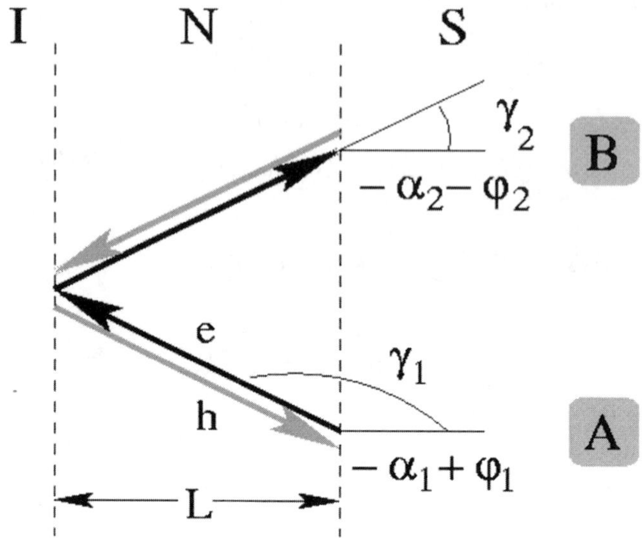

Figure 1. The example of the quasiparticle path corresponding to the Andreev reflections, giving a bound state. The quasiparticle is trapped in the normal region because of normal reflection at the I/NM surface and the Andreev reflection at the NM/SC interface. The total phase accumulated during one cycle is equal: $-(\alpha_1 + \alpha_2) \pm (\varphi_1 - \varphi_2) + \beta(E)$.

trajectory. The total accumulated phase in this case consists of contribution from Andreev reflections at point A: $-\alpha_1 + \varphi_1$ and B: $-\alpha_2 + \varphi_2$ as well as contribution from the propagation through the normal metal $\beta(E)$. $\alpha_{1(2)} = arccos(E/|\Delta_0|)$ is the Andreev reflection phase shift, while $\varphi_{1(2)}$ is the phase of the SC order parameter at point A (B). $\beta(E) = 2L(k_e - k_h) + \beta_0$ is the electron-hole dephasing factor and describes the phase acquired during the propagation through the normal region, where the first term corresponds to the ballistic motion and the second one to the reflection at the I/NM surface. L is the thickness of NM, and k_e (k_h) is the electron (hole) wave vector. Thus the Bohr-Sommerfeld quantization condition is:

$$-(\alpha_1 + \alpha_2) \pm (\varphi_1 - \varphi_2) + \beta(E) = 2n\pi \tag{7}$$

where the $\pm(\varphi_1 - \varphi_2)$ stands for the trajectories in the $\pm k_y$ (parallel to the interface) direction.

If there is no phase difference between points A and B in the Fig. 1 (for example NM/SC interface), the bound states always appear in pairs symmetrically positioned around the Fermi level because of the time reversal symmetry in the problem. Moreover, due to the fact that there is no difference between electrons and holes at the Fermi level ($\beta(E = 0) = 0$), there is no $E = 0$ solution. In other words, the bound states always emerge at finite energies.

The situation is quite different if there is a phase difference ($\varphi_1 - \varphi_2$) between points A and B (see Fig. 1). The example can be the interfaces with d-wave superconductors oriented in the (110) direction, where ($\varphi_1 - \varphi_2$) $= \pi$. In this case, due to the additional phase shift π the bound states can emerge even at zero energy. Such zero-energy Andreev bound states, in the case of high-T_c superconductors, have been predicted by Hu [20] and are known as *zero-energy mid-gap states.* The presence of the Andreev bound states at zero energy features in many important effects, like zero-bias conductance peaks, π-junction behavior, anomalous temperature dependence of the critical Josephson current, paramagnetic Meissner effect, time reversal symmetry breaking and spontaneous interface currents [17, 18].

Although the zero-energy states (ZES) are likely to appear when the phase of the order parameter at the interface is not constant, the resulting density of states at the Fermi energy is energetically unfavorable and any mechanism able to split these states will lower the energy of the system [18, 26]. On of these is the self-induced Doppler shift [27, 17] $\delta = ev_F A$, where A is a vector potential. The situation is schematically depicted in the Fig. 2. At low temperature ($T^* \approx (\xi_0/\lambda)T_c$, where λ

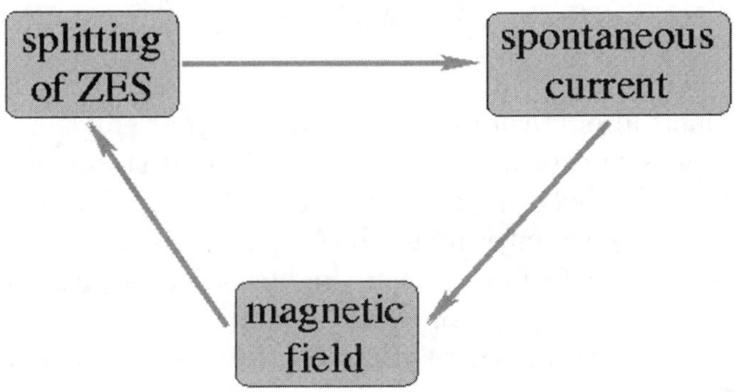

Figure 2. Generating of the spontaneous currents.

is the penetration depth of the magnetic field) the splitting of the zero energy states produces a surface current. This current generates a magnetic field (screened by a supercurrent), which further splits ZES due to the Doppler shift effect. The effect saturates when the magnetic energy is equal to the energy of the Doppler shifted ZES.

Naturally, the Andreev bound states also arises in $I/FM/SC$ heterostructures [21, 28–31, 23, 24, 32]. More importantly, as it was first predicted by Kuplevakhskii & Fal'ko [21], it is possible to shift these states to zero energy by tuning the exchange splitting. So the crossing of the zero energy solution can be obtained either by changing the phase difference $(\varphi_1 - \varphi_2)$ or by varying FM coherence length (exchange field).

The properties of such bound states have been also studied fully quantum-mechanically within lattice models of the FM/SC systems [31, 23, 24] and similar their behavior have been obtained. Interestingly, it turns out, that as in the case of the high-T_c structures [27], such zero energy Andreev states support spontaneous currents flowing in the ground state of the FM/SC system [22–24]. The mechanism of generating of such currents is the same, as earlier discussed, namely the self-induced Doppler shift. So in fact, when the current flows, such one of the states will be twice shifted: once due to the exchange (Zeeman) splitting, and the second time due to the Doppler shift.

For energies less than superconducting gap, the only Andreev bound states will contribute to the density of states $\rho(E)$. However, as it was mentioned, for fixed thickness and exchange splitting, there will be Andreev bound states at different energies, for different angles of particle incidence (γ_2 in the Fig. 1). Thus to get the density of states, one has to sum the energies of these states over all values of γ_2:

$$\rho(E) = \sum_{\gamma_2=-\pi/2}^{\pi/2} \delta(E - E_{bound}) \tag{8}$$

and talk, in fact, about Andreev bands rather that single states. However, all that was said on properties of the bound states, remains true for Andreev bands too. In particular the splitting of the whole band due to the spontaneous current is illustrated in the Fig. 3. The additional structure comes from the other (higher order) Andreev reflections. Superconducting energy gap $\Delta_0 = 0.376$ in this figure.

There is also a strong correlation between Andreev bound states (bands) and the pairing amplitude [31, 22, 23]. Each time the pairing amplitude at the I/FM interface changes its sign, the Andreev bound state (band) crosses the Fermi energy. Moreover in this case the spontaneous current is generated.

From the experimental point of view the density of states, in particular its temperature dependence, can be a good measure of the current carrying ground state. At certain thicknesses of FM for which the current flows there is a huge drop in the $\rho_{tot}(\varepsilon_F)$ at characteristic temperature $T^* \approx (\xi_S/\lambda)T_c$, where ξ_S and λ are coherence length and penetration

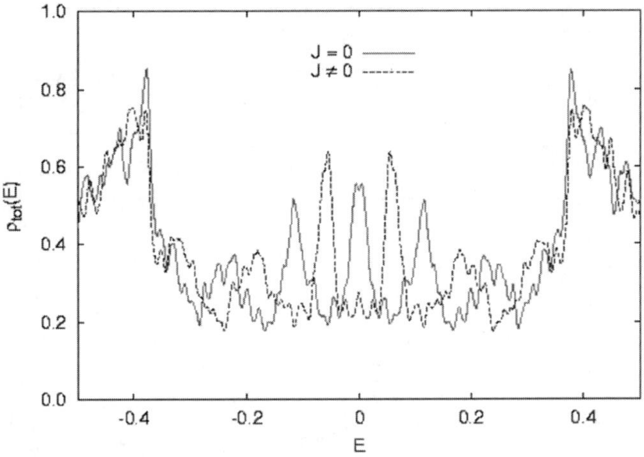

Figure 3. Doppler splitting of the zero-energy state. From Ref. [23].

depth respectively. T^* simply indicates the temperature at which magnetic instability, which leads to the generation of the current, takes the place. Such behavior is depicted in the Fig. 4 and should be observable experimentally. If there is no current the DOS is due to the Andreev band and is almost constant (we are well below T_c), and as soon as the current starts to flow the Andreev band splits so we observe a drop in $\rho_{tot}(\varepsilon_F)$. The important point is that T^* and T_c are different temperatures.

1.4 Spontaneous current

The most remarkable feature of our calculations is that the solution of the $SPHFG$ equations frequently converges to a solution with the finite current $j_y(n)$ even though the external vector potential is zero. The typical example of such a current, flowing parallel to the FM/SC interface, ($j_y^{tot}(n) = j_{y\uparrow}(n) + j_{y\downarrow}(n)$) is shown in the Fig. 5 for a few values of the exchange splitting. Behavior of the current, as a function of the layer index, is very similar to the density of states at the Fermi level. The oscillating nature of the current comes from the Friedel like oscillations of the DOS [23]. This is because current is proportional to the DOS at the Fermi level. Within semiclassical calculations, which neglect these effects the current is very smooth [32].

Another important issue is the distribution of the current through the whole trilayer structure. We find that it flows mostly in the positive y direction on ferromagnetic side and in the negative direction in the

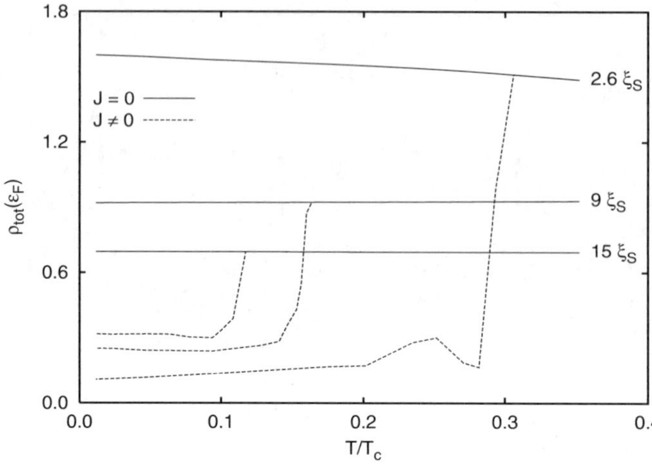

Figure 4. The temperature dependence of the surface (FM/vacuum) density of states at the Fermi energy for various thicknesses of the FM slab in the figure. The solid (dashed) line corresponds to the solution without (with) the current. From Ref. [24]

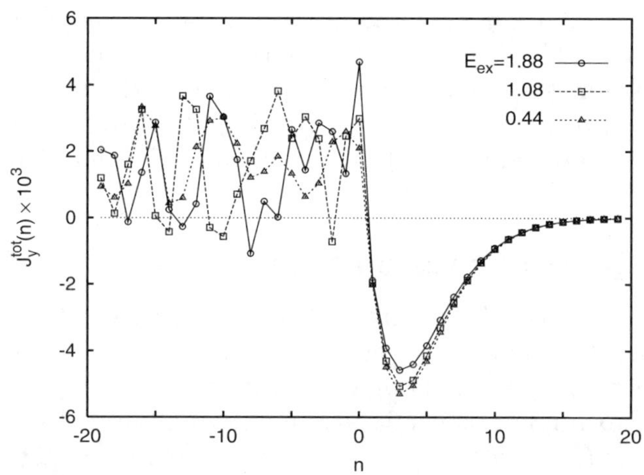

Figure 5. The total (spontaneous) current $j_y^{tot}(n) = j_{y\uparrow}(n) + j_{y\downarrow}(n)$ flowing parallel to the FM/SC interface for a number of exchange splittings. From Ref. [23].

superconductor. Notably the total current, integrated over the whole sample, is equal to zero within numerical accuracy. This is as it should be for the true ground state and found to be in the $FFLO$ state, where the current associated with the unpaired electrons is balanced by the supercurrent flowing in the opposite direction. Similarly here (see Fig. 6).

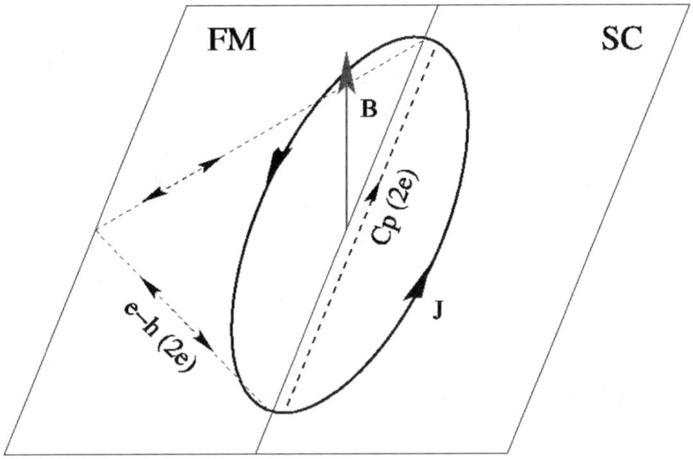

Figure 6. Schematic view of the current distribution.

Obviously, the spontaneous current distribution (see Fig. 6) generates the magnetic field through the sample. The total magnetic flux weakly depends on the thickness of the sample and the exchange splitting. Its magnitude is found to be a fraction of the flux quantum $\Phi_0 = h/2e$ and is smaller than upper critical field of the bulk superconductor. This is rather a large field and could be observable in temperature dependent measurements (see Fig. 7).

1.5 Transport properties

Some information on spontaneous currents can be also obtained from conductance calculations. To do so we attached a normal metal electrode to our FM/SC system and calculate current through $NM/FM/SC$ system using nonequilibrium Keldysh Green's function technique [33]. To calculate this current in terms of various physical processes we went along the way outlined in Ref. [34] and got corresponding spin polarized formula for the current as a sum of four different contributions $I = I_1 + I_2 + I_3 + I_A$, where:

$$\begin{aligned} I_1 &= 4\pi^2 t_{NF}^2 \frac{e}{\hbar} \sum_\sigma \int d\omega |1 \\ &+ G_{FN\sigma}^{11r}(\omega)|^2 \rho_{NN\sigma}^{11}(\omega) \rho_{FF\sigma}^{11}(\omega)[f(\omega - eV) - f(\omega)] \end{aligned} \tag{9}$$

$$\begin{aligned} I_2 &= 8\pi^2 t_{NF}^2 \frac{e}{\hbar} \sum_\sigma \int d\omega \mathrm{Re}\{t_{NF} G_{NF\sigma}^{21a}(\omega)[1 + G_{FN\sigma}^{11r}(\omega)]\} \\ &\times \rho_{NN\sigma}^{11}(\omega) \rho_{FF\sigma}^{12}(\omega)[f(\omega) - f(\omega - eV)] \end{aligned} \tag{10}$$

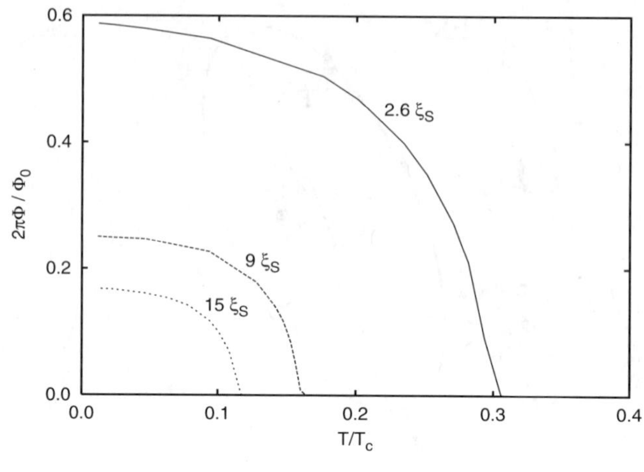

Figure 7. The temperature dependence of the total magnetic flux for thickness of the FM slab $L/\xi_S = 2.6$ (solid), 6 (dashed) and 15 (dotted curve).

$$I_3 = 4\pi^2 t_{NF}^4 \frac{e}{\hbar} \sum_\sigma \int d\omega |G_{FN\sigma}^{12}(\omega)|^2 \rho_{NN\sigma}^{11}(\omega)\rho_{FF-\sigma}^{22}(\omega)[f(\omega - eV) - f(\omega)] \quad (11)$$

$$I_A = 4\pi^2 t_{NF}^4 \frac{e}{\hbar} \sum_\sigma \int d\omega |G_{FF\sigma}^{12}(\omega)|^2 \rho_{NN\sigma}^{11}(\omega)\rho_{LL-\sigma}^{22}(\omega)[f(\omega - eV) - f(\omega + eV)] \quad (12)$$

I_1 corresponds to normal electron tunneling between electrodes, I_2 is a net transfer of single electron with creation or annihilation of pairs as an intermediate state. I_3 corresponds to a process in which electron from normal electron is converted to a hole in superconductor - branch crossing process in language of BTK theroy [35], while I_A is the Andreev tunneling.

The differential conductance $G(eV) = dI/d(eV)$ as a function of $eV = \mu_{NM} - \mu_{SC}$ is shown in the Fig. 8. Clearly, if there is a spontaneous current in the ground state, the conductance peak is split, similarly as in the DOS. We could expect such behavior because $G(eV)$ is proportional to the DOS at the Fermi energy. And again this effect could be observable in the tunneling experiments.

We have also extracted Andreev conductance form the total one and ploted in the Fig. 9. We can see that conductance associated with the Andreev processes is strongly enhanced when the current flows in the ground state. Unfortunately it could be very difficult experimentally

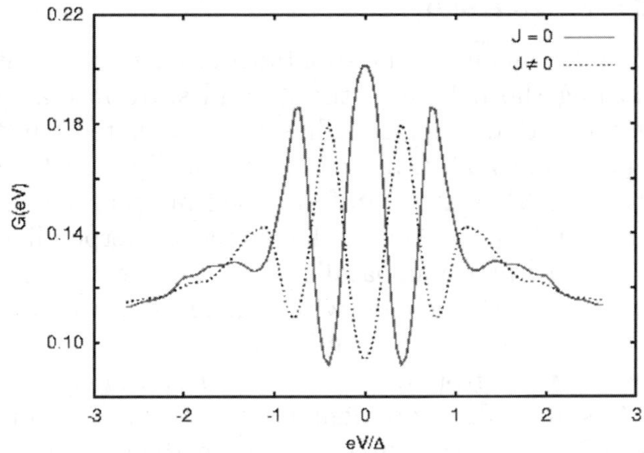

Figure 8. The total differential conductance for the solution with and without the spontaneous current.

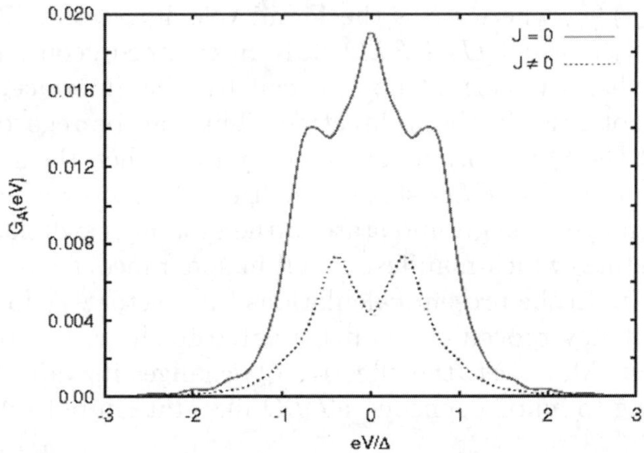

Figure 9. Corresponding Andreev differential conductance for the solution with and without the spontaneous current.

measure Andreev conductance only. Despite the fact that for energies less than SC gap the only allowed process is Andreev reflection, as in the point contact geometry, it doesn't work in our system. The problem is that even at very low energies there is a finite DOS at the Fermi level due to ferromagnet. Naturally the pairing amplitude is induced in FM slab but this is not true energy gap in the quasiparticle spectra and we always deal with some single electron processes in tunneling events.

1.6 2D FFLO state

Before closing discussion on the spontaneous current we wish to make a remark regarding the nature of the ground state in our system. To begin with we recall that recently it has been predicted [36] that under certain conditions a $3D$-$FFLO$ state is energetically more favorable than usual $1D$ state. The $3D$ state manifests itself in oscillatory behavior of the pairing amplitude not only in the direction perpendicular to the interface but also in direction parallel to it. Moreover, changing the thickness of the FM slab, one can switch the ground state of the system between $3D$ and $1D$-$FFLO$ state [36, 23].

The current carrying ground state of our system can be interpreted as a $2D$-$FFLO$ state. The argument is as follows: The oscillations of the pairing amplitude in the direction perpendicular to the interface occur regardless whether the spontaneous current flows or not. Within the $FFLO$ theory [7, 8], the period of the oscillations is related to the x-component of the center of mass momentum of the Cooper pair $\mathbf{Q}$. On the FM side of our model the $FFLO$ periodicity is governed by $\mathbf{Q} = (2E_{ex}/v_F)\frac{\mathbf{v}_F}{v_F}$, where $\mathbf{v}_F$ is the Fermi velocity vector. This can be interpreted as the usual $1D$-$FFLO$ state in confined geometry. On the other hand, when the current flows parallel to the interface, there is a finite vector potential in the y-direction. This can be regarded as a y-component of the $\mathbf{Q}$-vector. So one can say that when the spontaneous current flows, the $2D$-$FFLO$ state is realized. Moreover when the FM thickness is changed the ground state of the system is switched between $2D$- and $1D$-state, which manifests itself in spontaneous current flow or in the lack of it. In the present calculations this vector was found during the self-consistency procedure, as it is related to the vector potential in the y-direction. Moreover, the effective Q_y changes its value from layer to layer leading to inhomogeneous $FFLO$-like state in both dimensions.

1.7 Conclusions

The competition between ferromagnetism and superconductivity in FM/SC heterostructures give raise to the Fulde - Ferrell - Larkin - Ovchinnikov ($FFLO$) state in these systems. The original bulk $FFLO$ state manifests itself in a spatial oscillations of the SC order parameter as well as in spontaneously generated currents flowing in the ground state of the system. We have argued that a very interesting version of this phenomenon accures in FM/SC proximity systems. In short, due to the proximity effect and Andreev reflections at the FM/SC interface, the Andreev bound states appear in the quasiparticle spectrum. These states can be shifted to the zero energy by tuning the exchange splitting or the thickness of the ferromagnet, thus they became zero-energy midgap states which lead to various interesting effects. In particular, the occurence of spontaneous currents in the ground state can be related to the zero-energy states, as in the case of high-T_c superconductors. It seems that some combination of both phenomena is realized in a real systems. The fact that oscillatory behavior of SC order parameter is strongly correlated with the crossing of the Andreev bound states through Fermi energy and the generation of the spontaneous currents further support $FFLO$ - Andreev bound states picture. The experimental confirmation of the existence of the spontaneous (spin polarized) currents in the ground state would support the $FFLO$ - Andreev bound states scenario in these structures.

Acknowledgments

This work has been partially supported by the grant no. PBZ-MIN-008/P03/2003.
BLG would like to thank the Center for Computational Material Science (CMS) of TU Wien for hospitality during the preparation of the above talk.

References

[1] C. J. Lambert, R. Raimondi, J. Phys. Condens. Matter **10**, 901 (1998).

[2] A. F. Andreev, Sov. Phys. JETP **19**, 1228 (1964).

[3] N. F. Berk, J. R. Schrieffer, Phys. Rev. Lett. **17**, (1966) 433; C. Pfleiderer,M. Uhlarz, S. M. Hayden, R. Vollmer, H. v. Lohneysen, N. R. Bernhoeft, G. G. Lonzarich, Nature **412**, (2001) 58; D. Aoki, A. Huxley, E. Ressouche, D. Braithwaite, J. Flouquet, J. -P. Brison, E. Lhotel, C. Paulsen, Nature **413**, (2001) 613.

[4] G. E. W. Bauer, Yu. V. Nazarov, D. Huertas-Hernando, A. Brataas, K. Xia, P. J. Kelly, Materials Sci. Eng. B **84**, (2001) 31; S. Oh, D. Youm, M. R. Beasley, Appl. Phys. Lett. **71**, (1997) 2376; L. R. Tagirov, Phys. Rev. Lett. **83**, (1999) 2058.

[5] G. Blatter, V. B. Geshkenbein, L. B. Ioffe, Phys. Rev. B **63**, (2001) 174511.

[6] M. J. M. de Jong, C. W. J. Beenakker, Phys. Rev. Lett. **74**, 1657 (1995).

[7] P. Fulde, A. Ferrell, Phys. Rev. **135**, A550 (1964).

[8] A. Larkin, Y. Ovchinnikov, Sov. Phys. JETP **20**, 762 (1965).

[9] A. I. Buzdin, L. N. Bulaevskii, S. V. Panyukov, JETP Lett. **35**, 178 (1982); A. I. Buzdin, M. V. Kuprianov, JETP Lett. **52**, 487 (1990).

[10] Z. Radović, M. Ledvij, L. Dobrosavljević-Grujić, A. I. Buzdin, J. C. Clem, Phys. Rev. **B44**, 759 (1991).

[11] E. A. Demler, G. B. Arnold, M. R. Beasley, Phys. Rev. **B55**, 15 174 (1997).

[12] Y. N. Proshin, M. G. Khusainov, JETP Lett. **66**, 562 (1997); M. G. Khusainov, Y. N. Proshin, Phys. Rev. **B56**, R14283 (1997); *ibid.* **62**, 6832 (2000).

[13] H. K. Wong, B. Y. Jin, H. Q. Yang, J. B. Ketterson, J. E. Hilliard, J. Low Temp. Phys. **63**, 307 (1986).

[14] T. Kontos, M. Aprili, J. Lesueur, X. Grison, Phys. Rev. Lett. **86**, 304 (2001).

[15] L. N. Bulaevksii, V. V. Kuzii, A. A. Sobyanin, JETP. Lett. **25**, 290 (1977).

[16] M. Sigrist, T. M. Rice, J. Phys. Soc. Japan **61**, 4283 (1992).

[17] M. Sigrist, T. M. Rice, Rev. Mod. Phys. **67**, 503 (1995); S. Kashiwaya and Y. Tanaka, Rep. Prog. Phys. **63**, 1641 (2000); T. Löfwander, V. S. Shumeiko, G. Wendin, Supercond. Sci. Technol. **14**, R53 (2001).

[18] M. Sigrist, Prog. Theor. Phys. **99**, 899 (1998).

[19] P. G. de Gennes and D. Saint-James, Phys. Lett. **4**, 151 (1963).

[20] C. Hu, Phys. Rev. Lett. **72**, 1526 (1994).

[21] S. V. Kuplevakhskii, I. I. Fal'ko, JETP Lett. **52**, 340 (1990).

[22] M. Krawiec, B. L. Györffy, J. F. Annett, Phys. Rev. **B66**, 172505 (2002).

[23] M. Krawiec, B. L. Györffy, J. F. Annett, European Phys. J. **B32**, 163 (2003).

[24] M. Krawiec, B. L. Györffy, J. F. Annett, Physica **C387**, 7 (2003).

[25] K. P. Dunkan, B. L. Györffy, Ann. Phys. (NY) **298**, 273 (2002).

[26] M. Fogelström, S. -K. Yip, Phys. Rev. **B57**, R14 060 (1998).

[27] S. Higashitani, J. Phys. Soc. Jpn. **66**, 2556 (1997).

[28] A. Kadigrobov, R. I. Shekhter, M. Jonson, Z. G. Ivanov, Phys. Rev. **B60**, 14 593 (1999).

[29] M. Zareyan, W. Belzig, Y. V. Nazarov, Phys. Rev. Lett. **86**, 308 (2001).

[30] N. M. Chtchelatchev, W. Belzig, Y. V. Nazarov, C. Bruder, JETP Lett. **74**, 323 (2001).

[31] E. Vecino, A. Martin-Rodero, A. L. Yeyati, Phys. Rev. **B64** 184502 (2001).

[32] M. Krawiec, B. L. Györffy, J. F. Annett, cond-mat/0302162.

[33] L. V. Keldysh, Sov. Phys. JETP **20**, 1018 (1965).

[34] J. C. Cuevas, A. Martin-Rodero, A. L. Yeyati, Phys. Rev. **B54**, 7366 (1996).

[35] G. E. Blonder, M. Tinkham, T. M. Klapwijk, Phys. Rev. **B25**, 4515 (1982).

[36] Y. A. Izyumov, Y. N. Proshin, M. G. Khusainov, Phys. Usp. **45**, (2002) 109; Y. A. Izyumov, Y. N. Proshin, M. G. Khusainov, JETP Lett. **71**, (2000) 138; M. G. Khusainov, Y. A. Izyumov, Y. N. Proshin, Physica B **284-288**, (2000) 503.

EXCHANGE FORCE IMAGE OF MAGNETIC SURFACES

— *A First-Principles Study on NiO(001) Surface* —

Hiroyoshi Momida*, Tamio Oguchi

ADSM, Hiroshima University
1-3-1 Kagamiyama, Higashihiroshima 739-8530
Japan

Abstract A recently proposed surface atom-probe technique, exchange force microscopy, is examined on an antiferromagnetic NiO (001) surface system. It is shown that atomic force of a ferromagnetic Fe probe on the surface gives a clear spin image when the probe is located within 1Å above the contact point. Exchange force images show antiferromagnetic pattern of the Ni sites along the [110] direction and asymmetry around the O sites. The asymmetric feature comes from the superexchange interaction between the probe and the second-layer Ni atoms via the surface O ion, being a key proof of the exchange force image on observation.

Keywords: Exchange force microscopy, surface magnetism, first-principles calculation.

2.1 Introduction

Exchange force microscopy (EFM) is a spin-dependent extension of non-contact type atomic force microscopy (AFM) by measuring forces acting on a ferromagnetic probe depending on relative spin orientation to the surface atoms [1]. The basic setup of EFM is quite analogous to that of magnetic force microscope (MFM). If the tip and surface are close to each other, typically the order of Å, short-range exchange interaction becomes dominant over relatively long-range magnetic dipole one and a spin image of the magnetic surface can be observed in an atomic scale.

*present address: National Institute for Materials Science, 1-2-1 Sengen, Tsukuba 305-0047, Japan

S. Halilov (ed.), Physics of Spin in Solids: Materials, Methods and Applications, 17–24.

A lot of efforts have been made to development of EFM instrumental setup and surface preparation [2–4] as well as theoretical predictions and quantitative estimation of EFM from first principles [5, 6]. Nakamura has calculated atomic forces between Fe thin films with two kinds of relative magnetization orientations, parallel and anti-parallel, to evaluate exchange force in a realistic system [5]. The magnitude of obtained exchange force is an order of nN, which can be measured with the present non-contact AFM techniques. He found that the exchange force shows an oscillatory behavior as a function of the film distance and that the interaction is dominated by direct exchange between the neighboring d orbitals for short distances while becomes indirect or RKKY-type via more extended s and p orbitals for larger distances. ¿From the lateral variation of the exchange force and its derivative with respect to the distance, it has been concluded that the exchange force may provide us clear information of the spin density distribution of the magnetic surface [6].

However, it may be hard to flip the spin direction of the probe during the observation and more essentially the exchange force cannot be extracted precisely from the atomic force because both atomic and exchange forces have the same symmetry on the ferromagnetic surface. Instead of measuring the ferromagnetic surface with a spin-flipped tip, it should be much easier to measure an anti-ferromagnetic surface with a spin-fixed ferromagnetic tip. This is a quite similar idea to spin-polarized scanning tunneling microscopy (STM) measurement for anti-ferromagnetic Mn adsorbed on W(110) reported recently [7].

Quite recently we have carried out first-principles calculations for type-II anti-ferromagnetic NiO (001) clean surface to investigate surface structure, electronic structure and spin-density distribution [8]. We have found that the oxygen atoms at the surface may have the spin moment of $0.07\mu_B$, which is parallel to that of the Ni atom beneath. In this study, we calculate exchange force of a ferromagnetic Fe probe on antiferromagnetic NiO (001) surface to predict exchange force images of the magnetic surface in an atom scale.

2.2 Models and Methods

We calculate atomic forces of a ferromagnetic Fe probe on antiferromagnetic NiO (001) surface to evaluate exchange forces. An important point here is that in case of anti-ferromagnetic surface, we can obtain exchange force with the ferromagnetic tip just by subtracting the forces on antiferromagnetically inequivalent sites. To model such a system, Fe mono-layer on 5-layer NiO (001) slab is adopted for the EFM calculations as shown in Fig. 1.

Figure 1. Model system of exchange force microscopy: ferromagnetic Fe monolayer on 5-layer antiferromagnetic NiO(001) slab. Contour plot represents spin density distribution on (100) cross-section. Solid (broken) lines represent positive (negative) spin density. Interval of the density is 0.002 μ_B/a_0^3, where a_0 is the Bohr radius. High spin density region around Ni and Fe is not shown.

For comparison, we have performed electronic structure calculations for clean NiO (001) surface with use of a 9-layer slab model as well as bulk NiO. We use lattice constant of 4.084 Å taken from the equilibrium value of bulk NiO. In the present work, we assume no structural relaxation in case of the EFM calculations. We especially focus on probe-height and lateral dependences of the exchange force. The lateral dependence gives us the exchange force images we like to investigate. Our calculations are based on local-spin-density approximation to the density functional theory. One-electron scalar-relativistic Kohn-Sham equations are solved self-consistently by using full-potential linear augmented plane wave (FLAPW) method. The Kohn-Sham wavefunctions are expanded by the basis set with the energy cutoff of 15 Ry. The Brillouin zone integration is done by using two-dimensional uniform **k**-mesh including six **k** points in the irreducible zone.

2.3 Results and Discussion

Electronic Structure and Spin Density in Bulk NiO

Before showing results for NiO surface, let us summarize the electronic structure and spin density in bulk NiO, which have been well studied for a couple of decades. Figure 2(c) shows partial density of states (DOS) calculated for bulk antiferromagnetic NiO, which is a reference to the surface system we like to study.

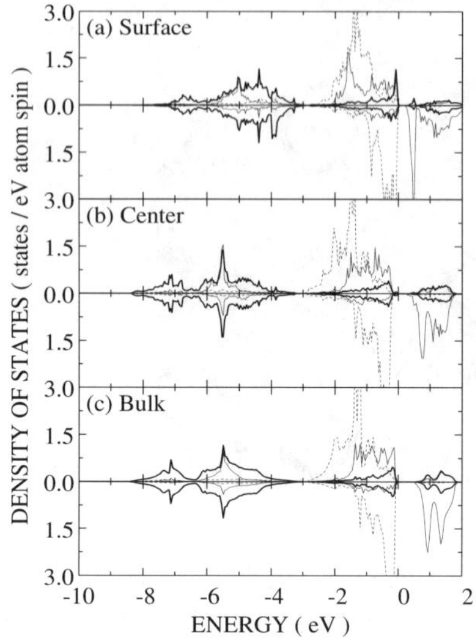

Figure 2. Calculated partial density of states of (a) surface layer and (b) center layer in 9-layer NiO(001) slab and (c) bulk NiO. Thick solid, thin solid and broken lines denote O-p and Ni-d e_g and t_{2g} components, respectively. The energy zero is taken at the top of the valence band of each system. Top and bottom panels represent the majority and minority spin states, respectively, of each layer.

Spin magnetic moment at the Ni site is $1.16\mu_B$, which comes from hole in Ni-d e_g ($3z^2 - r^2$ and $x^2 - y^2$) minority-spin band. So, the spin-density distribution around Ni reflects the shape of the e_g orbital, as equivalently shown in the spin density distribution in the center layer of the NiO slab in Fig. 3.

There is strong hybridization between O-p and Ni-d e_g orbitals and certain hole amplitude may exist also at the O sites. One can see weak spin-density distribution around O in Fig. 3 but its integrated quantity, namely spin moment, becomes zero at the O sites by symmetry.

Electronic Structure and Spin Density in NiO (001) Surface

Contrast to the bulk NiO, the O ions may have finite size of spin moment at surface because of symmetry breaking. Actually we get such finite spin moment at surface O sites as much as $0.07\mu_B$. The surface spin moment is parallel to that of the Ni site underneath and originates in O-p_z orbital as shown in Fig. 3. The O spin moments are diminished

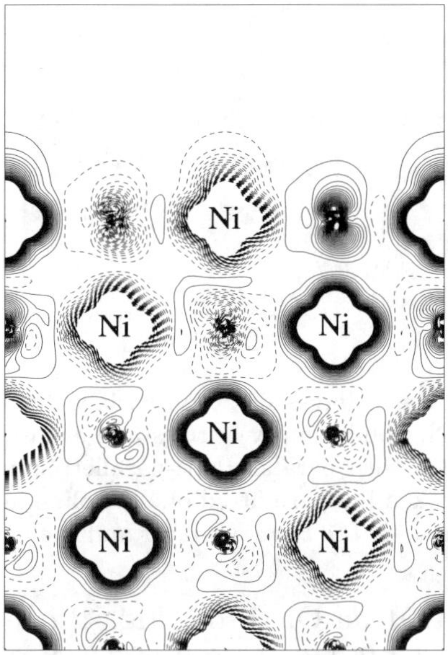

Figure 3. Calculated spin density distribution on (100) cross-section in 9-layer NiO (001) slab. Solid (broken) lines represent positive (negative) spin density. Interval of the density is 0.002 μ_B/a_0^3, where a_0 is the Bohr radius. High spin density region around Ni is not shown.

quickly beyond the second layer, approaching to the bulk value. On the other hand, the spin density distribution around Ni and the spin moments are very rigid in nature.

Figure 2 shows partial DOS of the surface and center layer. Calculate DOS of the center layer is just like bulk DOS while one can see a reduction of the band gap at surface due to Ni-d e_g orbital. This reduction comes from lack of hybridization of the e_g orbital vertical to the surface ($3z^2 - r^2$ component) while surface-parallel ($x^2 - y^2$) component of the e_g orbital is still bulk-like. Small spin polarization at the O sites can be recognized, mostly originating from O-p_z orbital.

Atomic and Exchange Forces of Fe Probe on NiO(001) Surface

Calculated atomic forces are smooth functions of the probe height as shown in Fig. 4.

The probe height is taken as the distance between the Fe mono-layer and the surface layer of NiO (001) (see Fig. 1). The Fe probe atoms are attracted more strongly by O than by Ni. But exchange force, which is defined as the atomic force difference between different spin orientations

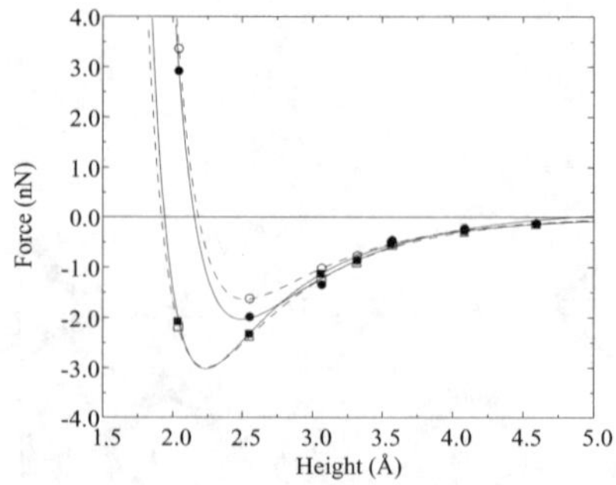

Figure 4. Calculated atomic forces of Fe on surface Ni and O sites in antiferromagnetic NiO (001) in nN. Solid (empty) circles and squares denote atomic force of Fe atom on Ni and O sites with parallel (anti-parallel) spin moment to that of Fe, respectively. Solid and broken lines are fitted one to spline functions. Probe height is defined as the distance between the Fe mono-layer and the surface layer of NiO (001).

is larger on Ni than that on O. ¿From Fig. 4, one can see the contact points at the heights of 1.9 Å on O and of 2.2 Å on Ni.

Exchange force is defined as

$$F_{ex} = F_{up} - F_{down}, \tag{1}$$

where F_{up} and F_{down} are atomic forces of Fe probe with up and down spin moment, respectively.

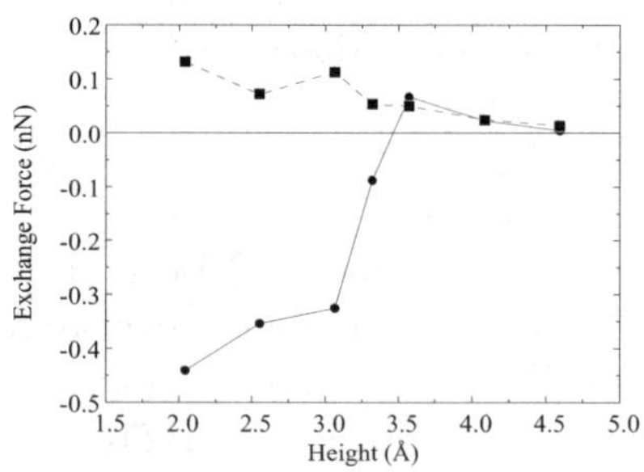

Figure 5. Probe-height dependence of exchange force on surface Ni and O sites in antiferromagnetic NiO (001) surface in nN. Solid circles and squares denote the exchange force of Fe on Ni and O sites.

Figure 5 shows strong ferromagnetic forces obtained on the Ni sites up to the probe height of 3.4 Å, which is about 1.2 Å above the contact point. The forces are changed to anti-ferromagnetic beyond that height while always antiferromagnetic on the O sites. The strong ferromagnetic forces may come from direct exchange between the Fe probe and Ni surface atoms while the antiferromagnetic forces from indirect or superexchange mechanism. The magnitude of the exchange force is the order of one tenth of nN, which should be observable with the state-of-the-art non-contact AFM techniques. We can expect atomic-scale resolution in the exchange force image below the height of 3.4 Å. Such a precise height control of the order of Å is also possible with the latest non-contact AFM techniques.

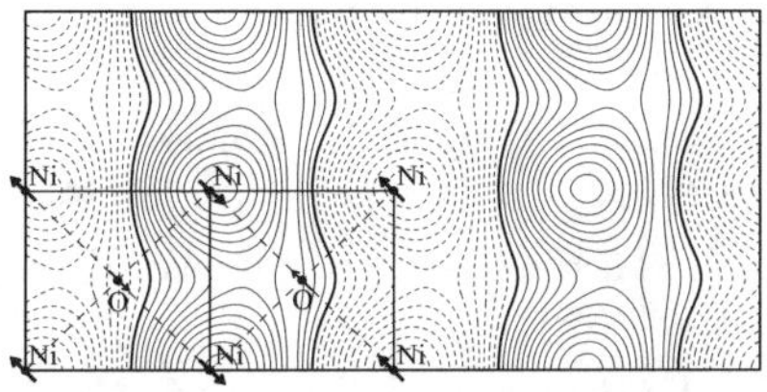

Figure 6. Lateral dependence of exchange force on surface Ni and O sites in antiferromagnetic NiO (001) surface. Solid and broken lines represent positive and negative exchange forces. Interval of the force is 0.04nN.

Figure 6 shows the lateral distribution of the exchange force, namely EFM image, of NiO (001) surface at a probe height near the contact point. Qualitatively equivalent but weaker contrast image is obtained at 1 Å above the contact point. Generally, the image tells us antiferromagnetic pattern of the NiO (001) surface along the [110] direction. However, finite exchange force on the O sites makes the image asymmetric. Since the charge density should be symmetric with respect to the surface oxygen sites, observing such asymmetry should be a direct proof of the exchange force.

2.4 Conclusions

We have performed first-principles electronic structure calculations for anti-ferromagnetic NiO (001) clean surface and Fe probe on it. We have found that exchange force images of NiO should be observable in an atomic scale and asymmetric image due to super-exchange via the surface O sites is a crucial proof of the exchange force.

Acknowledgments

The authors thank K. Mukasa, K. Nakamura, and H. Hosoi for their continuing encouragement. The computations have been done at the Supercomputer Center, ISSP, University of Tokyo.

References

[1] For recent reviews, see: S. Morita, R. Wiesendanger and E. Meyer (Eds.), *Noncontact Atomic Force Microscopy* (Springer-Verlag, Berlin, 2002); E. Meyer, H.J. Hug and R. Bennewitz, *Scanning Probe Microscopy* (Springer-Verlag, Berlin, 2004).

[2] K. Mukasa, K. Sueoka, H. Hasegawa, Y. Tazuke and K. Hayakawa, Mater. Sci. Eng. B **31**, 69 (1995).

[3] H. Hosoi, K. Sueoka, K. Hayakawa and K. Mukasa, Appl. Surf. Sci. **157**, 218 (2000).

[4] H. Hosoi, M. Kimura, K. Hayakawa, K. Sueoka and K. Mukasa, Appl. Phys. A**72** Suppl., S23 (2001).

[5] K. Nakamura, H. Hasegawa, T. Oguchi, K. Sueoka, K. Hayakawa and K. Mukasa, Phys. Rev. B **56**, 3218 (1997).

[6] K. Nakamura, T. Oguchi, H. Hasegawa, K. Sueoka, K. Hayakawa and K. Mukasa, Appl. Surf. Sci. **142**, 433 (1999).

[7] S. Heinze, M. Bode, A. Kubetzka, O. Pietzsch, X. Nie, S. Blügel and R. Wiesendanger, Science **288**, 1805 (2000).

[8] H. Momida and T. Oguchi, J. Phys. Soc. Jpn. **72**, 588 (2003).

SPIN-DEPENDENT TUNNEL CURRENTS FOR METALS OR SUPERCONDUCTORS WITH CHARGE-DENSITY WAVES

A. M. Gabovich, A. I. Voitenko
Institute of Physics, prospekt Nauki 46, 03028 Kiev-28, Ukraine

Mai Suan Li, H. Szymczak
Institute of Physics, Al. Lotnikow 32/46, PL-02-668 Warsaw, Poland

M. Pekala
Department of Chemistry, University of Warsaw, Al. Zwirki i Wigury 101, PL-02-089 Warsaw, Poland

Abstract We suggest to extend the well-known method of Tedrow and Meservey to investigate spin polarization P in ferromagnets. Namely, metals and superconductors partially gapped by charge-density waves (CDWs) are proposed as counter-electrodes instead of ordinary superconductors. Differential conductances $G(V)$ for the quasiparticle tunnel currents in external magnetic fields are calculated. The results are substantially different from those for ordinary superconductors. In particular, current-voltage characteristics are nonsymmetrical even for $P = 0$.

Keywords: Spin-dependent tunneling, charge-density waves, superconductors, magnetic field, spin polarization, current-voltage characteristics

3.1 Introduction

Spin-polarized electron tunneling between superconductor (S) and ferromagnetic (FM) electrodes is a powerful method for studying both the electron properties of the paired state and the spin-splitted band structure of the itinerant electron spectrum [1, 2]. One of the main tasks here consists in the determination of the electron polarization P inside the

S. Halilov (ed.), Physics of Spin in Solids: Materials, Methods and Applications, 25–42.

ferromagnet which is defined as

$$P = \frac{N_{FM\downarrow} - N_{FM\uparrow}}{N_{FM\downarrow} + N_{FM\uparrow}}, \tag{1}$$

where $N_{FM\downarrow(\uparrow)}$ is the density of states (DOS) of the "majority" (" minority") electrons with spins directed opposite to (along) the direction of the magnetic field H. At the same time, the corresponding "majority" magnetic moments μ_s are directed *along* H, since $\mu_s = -\mu_B < 0$. Here μ_B is the Bohr magneton. The definition (1) is not unique, and transport properties may be better described by other combinations of majority and minority current contributions [3]. A proper choice is crucially important for calculations in specific cases, but for the problem discussed all changes might be reduced to the free parameter P renormalization.

The remarkable idea of Tedrow and Meservey [4, 5] consists in the estimation of P through the values of the differential tunnel conductivity $G(V) \equiv dJ/dV$ measured at definite voltages V and magnitudes of the external magnetic field H applied to the junction [1, 5]. Here J is a tunnel current. The method should work in this S-I-FM (superconductor-insulator-ferromagnet) set-up because the initially identical peaks of conductivities $G_\uparrow(V)$ and $G_\downarrow(V)$ from both spin subbands shift due to the Zeeman effect in the superconducting films when the field is switched on [6] and their amplitudes deform downwards and upwards nonsymmetrically.

Unfortunately, the application of this scheme, promising in principle, led for the junctions Al-Al_2O_3-FM, with FM = Ni and Co, to the deduced P of the wrong positive sign (i. e. the majority of the magnetic moments of tunneling electrons were found to be in the field direction), whereas the band calculations predicted that the minority-spin electrons should give the prevailing contribution to the DOS at the Fermi energy level and, hence, to the overall current [1, 7]. To solve the apparent controversy, a number of theoretical studies were carried out changing the starting naive picture of the tunneling process. First, it was recognized that the tunneling spin-splitted DOSes for ferromagnets differ from the band ones because the probability of the electron penetration into the barrier region depends on the kind of intermediate electronic states involved [1, 7–9]. The second required modification makes allowance for the non-Ohmic (Fowler-Nordheim) character of conductivity caused by the electric field distortion of the primordial barrier's rectangular shape [10]. Finally, the Zeeman splitting of the $G(V)$ peak in the superconducting electrode is drastically diminished by the spin-orbit interactions especially effective for heavy elements, with the respective scattering rate proportional to Z^4, where Z is the atomic number [1, 11]. Broadly speaking, the modern approaches treat the whole junction as a single entity and takes into account the interface states and possible structural disordering [7, 12].

The spin mechanism of the superconductivity suppression [6, 13–15], with the discussed spin splitting of $G(V)$-dependences being its precursor, can dominate over the orbital (Meissner) depairing [16, 17] only in special situations. For example, it can occur in thin film superconducting electrodes of the Al-Al_2O_3-FM sandwiches with the magnetic field parallel to the junction plane, since the orbital depairing is small for thin enough films and small mean free path l [1, 17, 18].

In the general case all the listed factors act simultaneously and their interplay is rather complicated. Hence, it becomes clear that the resources for selecting proper superconducting covers are not very numerous. At the same time, the use of the paramagnetic effect in nonmagnetic electrodes to probe the ferromagnetic properties of the counterelectrodes seems quite helpful. Therefore, we propose a new class of tunneling partners for the ferromagnetic materials, namely, metals partially gapped by charge-density waves (CDWs) – CDWMs. [19–24]. So, the tunneling scheme now has the form CDWM-I-FM. An external magnetic field stimulates a paramagnetic effect analogous to that in superconductors [25–27]. On the other hand, the giant diamagnetic (Meissner) response does not appear for CDWMs at all because this state lacks for superfluid properties [28, 29]. As for the spin-orbit coupling, which leads to harmful spin-flips [11], its role can be diminished by an adequate choice of the light-atom constituents for CDW materials. But in any case, since the critical temperature T_d of the CDW transition usually is much larger than its superconducting counterpart T_c and the same remains true for the corresponding order parameters Σ and Δ (energy gaps $|\Sigma|$ and $|\Delta|$), a much larger Zeeman splitting can be obtained for CDW metals in comparison to that in superconductors, so that the spin-orbital smearing would not suppress totally the separation between $G_\uparrow(V)$ and $G_\downarrow(V)$ peaks.

3.2 Formulation

Below we analyse current-voltage characteristics (CVCs) for a tunnel junction between FM and CDW superconductor (CDWS), the latter including CDW metal as a particular case, more simple from the mathematical as well as the physical points of view. Nevertheless, the main emphasis will be placed on sandwiches with the normal CDWM as one of its covers. This case is practically more important and easier to examine. The bias voltage V is chosen as the difference between voltages at the itinerant (Stoner) ferromagnet and CDW superconductor: $V \equiv V_{\rm FM} - V_{\rm CDWS}$.

It is presumed that for H high enough to produce experimentally resolved splitting of the electron DOS peaks all domains inside the ferromagnet are completely aligned in the field direction [1]. We also anticipate that the bulk polarization is preserved during the tunneling process, i.e. the influence of the ferromagnet-insulator interface on the tunnel current is totally neglected. We fully recognize that, generally speaking, such is not the case, the boundary and disorder effects being very important [2, 3, 7, 12, 30–33]. However, taking into account these complications may be postponed until the specific CDWS(CDWM)-I-FM junctions are produced. The main goal of this publication is to consider the very possibility of the new type of counter-electrodes in tunnel junctions to study magnetic materials.

The properties of the partially-gapped CDWS electrode are characterized in the framework of the Bilbro-McMillan model [24, 34]. According to this approach, which with an equal success describes both the Peierls insulating state in quasi-one-dimensional substances [19] and the excitonic insulating state in semimetals [29, 35], the Fermi surface (FS) consists of three sections. Two of them ($i = 1, 2$) are nested, with the corresponding fermion quasiparticle spectrum branches obeying an equation

$$\xi_1(\mathbf{p}) = -\xi_2(\mathbf{p} + \mathbf{Q}), \tag{2}$$

where $\mathbf{Q}$ is the CDW vector. So, the electron spectra here become degenerate (d) and a CDW-related order parameter appears. The rest of the FS ($i = 3$) remains undistorted under the electron-phonon (the Peierls insulator) or Coulomb (excitonic insulator) interaction and is described by the non-degenerate (nd) spectrum branch $\xi_3(\mathbf{p})$. A single superconducting order parameter Δ exists on the whole FS, whereas a dielectric (CDW) order parameter Σ appears only on the nested FS sections.

The resulting phase determined by the coupled superconducting $\Delta_{im}^{\alpha\gamma}$ and dielectric $\Sigma_{im}^{\alpha\gamma}$ matrix order parameters in the presence of the external magnetic field H without making allowance for the Meissner diamagnetism is described by a certain system of the Dyson-Gor'kov equations for the normal $\mathcal{G}_{ij}$ and anomalous $\mathcal{F}_{ij}$ temperature Green's functions [36]. Here Latin subscripts correspond to the section space, while Greek superscripts reflect the spin structure of the order parameters. The neglect of the diamagnetic effects when $\Delta_{im}^{\alpha\gamma} \neq 0$ is justified only for $H_p \ll H_{c2}$, where H_p is the paramagnetic limit [6, 13–15] and H_{c2} is the upper critical magnetic field [37] (hereafter we suggest that all possible CDW superconductors are of the II kind as is true at least for all known

CDW alloys and compounds). It should be also born in mind that in the mixed phase the diamagnetic response of the degenerate FS sections is smaller in the ratio of $\Delta^2/\left(\Delta^2+\Sigma^2\right)$ as compared with that appropriate to the Bardeen-Cooper-Scrieffer (BCS) superconductor possessing the same energy gap Δ [38, 39].

The orbital influence of the magnetic field on CDWs being not so large as in superconductors, nevertheless, can exist, at least in principle. Namely, if the nesting conditions are imperfect (which is always the case) and the Zeeman-splitting effects are negligible, a transverse magnetic field, which reduces the quasiparticle spectrum dimensionality, results in an increase of T_d. (It is also true for the critical temperature T_N of the spin-density-wave (SDW) state [40–42]. Moreover, field-induced SDWs were predicted [40, 43–45] and observed for organic substances [46, 47]. The situation for CDWs is more complicated, since in that case the magnetic field acts not only diamagnetically but also paramagnetically [26, 27].) But for present purposes, as it is clear from the aforesaid, one can disregard this effect while investigating the spin-splitted peaks of the differential conductivity for normal metals with CDW distortions. Of course, it does not mean that T_d itself does not depend on H if one goes beyond the approximation adopted in this publication. Since theoretical analysis of orbital and Pauli terms may lead to ambiguous results for $T_d(H)$ or $\Sigma(H)$, it is more useful to look at the available experimental data.

For a majority of CDW substances T_d is of the order of hundreds Kelvins [19, 20, 24] (in $SmTe_3$ $T_d \approx 1300\,\text{K}$ is even substantially higher than the melting temperature 1096 K [48]) and, as a consequence, the magnetic fields necessary to conspicuously alter T_d are inaccessible to experimentalists. There are, however, several compounds with smaller T_d, for which both the DOS spin-splittings and the dependences $T_d(H)$ can be observed relatively easily. First of all, the A15 compound V_3Si with $T_d(H=0) = 20.15\,\text{K}$ should be mentioned. Its investigation in the magnetic field showed [49] that the field-unduced CDW suppression $\Delta T_d \propto -H^2$ and is quite small indeed: even for a very large $H = 156\,\text{kOe}$ the correction was 0.6 K. Organic substances α-$(ET)_2MHg(SCN)_4$ (M = K, Tl, Rb, etc.) with $T_d = 8-10\,\text{K}$ (at the pressure $p = 1$ bar and $H = 0$) constitute another promising class of CDW objects [27, 50–52]. There is even a point of view [53, 54] that the diamagnetic orbital response in these compounds is connected to nonequilibrium persistent currents. One should also mention a Peierls quasi-one-dimensional metal $Per_2[Au(mnt)_2]$ ("Per" and "mnt" mean perylene and maleonitriledithiolate) with $T_d(H=0) = 12.2\,\text{K}$ and a similar quadratic decrease of T_d with H as in V_3Si [55].

Thus, while studying Pauli paramagnetic splitting in normal CDW metals no restrictions from above appear on the H amplitude other than the natural limit $\mu_B^* H < |\Sigma|$, where $|\Sigma|$ is the magnitude of the CDW order parameter. This inequality represents the paramagnetic limit for a CDW metal [25, 26, 50, 56] that in the first approximation has the same form and similar origin as its counterpart for superconductors. The Pauli paramagnetic suppression of the CDW order parameter is due to the fact that such a kind of the electron-hole pairing couples the bands (in the excitonic insulator) or the different parts of the one-dimensional self-congruent band (in the Peierls insulator) with the same spin direction, contrary to the SDW case, where current carriers with the opposite spin directions are paired. When the magnetic field is switched on, both congruent FS sections having the chosen spin projection shift either up or down in energy. Therefore, the nesting CDW vectors $\mathbf{Q}_{\downarrow,\uparrow}$ do not coincide any more, and the initial CDW state is gradually destroyed. It is remarkable that the CDW instability favours superconductivity in the mixed phase for $H \neq 0$ because the magnetic energy must overcome both CDW and superconducting energetical benefits. The enhancement of the paramagnetic limit in the CDW superconductor was predicted some time ago [36], although have not yet been confirmed experimentally.

In the conjectured absence of the orbital magnetism the thermodynamics of the CDW superconducting or normal metal in the magnetic field [36] is similar to the behavior of the BCS superconductor, where the diamagnetic phase with the homogeneous Δ and the initial T_c survives for a low enough H until the I-kind field-induced transition into the normal state occurs when H reaches the Clogston-Chandrasekhar value [37]. Since, we are going to deal with smaller fields, the intriguing problem of the nonhomogeneous state [25] analogous to the Larkin-Ovchinnikov-Fulde-Ferrel one in ordinary superconductors for $H \geq H_p$ will be not touched upon. Hence, making allowance for the spin-singlet structure (s-wave superconductivity and CDWs) of the matrix normal $\Sigma_{ij}^{\alpha\beta} = \Sigma\delta_{\alpha\beta}$ and anomalous $\Delta_{ij}^{\alpha\beta} = \mathbf{I}^{\alpha\beta}\Delta$ ($(\mathbf{I}^{\alpha\beta})^2 = -\delta_{\alpha\beta}$) self-energy parts in the weak coupling limit, we must consider the self-consistent equation system for the order parameters Σ and Δ in the case of $H = 0$. The explicit form of the equations can be found elsewhere [57].

Making use of the self-consistent solutions for functions $\Sigma(T)$ and $\Delta(T)$ we calculate a quasiparticle tunnel current $J(V)$ between a ferromagnet and a CDW superconductor (or a CDW normal metal when $\Delta \equiv 0$) according to the expressions which can be straightforwardly obtained by the Green's function method of Larkin and Ovchinnikov developed for BCS superconductors [58]. The particular case of $P = 0$ and $H = 0$ was treated in our previous publications, which contain all technical details [59–62]. Generally, the current $J(V)$ consists of six components:

$$J(V) = \sum_{\substack{f=nd,d,ib \\ s=\downarrow,\uparrow}} J_{f,s}(V); \tag{3}$$

$$J_{nd,\downarrow(\uparrow)} = \frac{(1-\mu)(1 \pm P)}{4eR} \int_{-\infty}^{\infty} d\omega\ K(\omega, V, T)\, |\omega \mp \mu_B^* H|\, f_{\mp}(\omega, H, \Delta); \tag{4}$$

$$J_{d,\downarrow(\uparrow)} = \frac{\mu(1 \pm P)}{4eR} \int_{-\infty}^{\infty} d\omega\ K(\omega, V, T)\, |\omega \mp \mu_B^* H|\, f_{\mp}(\omega, H, D); \tag{5}$$

$$J_{ib,\downarrow(\uparrow)} = \frac{\mu(1 \pm P)\Sigma}{4eR} \int_{-\infty}^{\infty} d\omega\ K(\omega, V, T)\mathrm{sgn}\,(\omega \mp \mu_B^* H)\, f_{\mp}(\omega, H, D). \tag{6}$$

Here

$$K(\omega, V, T) = \tanh\frac{\omega}{2T} - \tanh\frac{\omega - eV}{2T}, \tag{7}$$

$$f_{\mp}(\omega, H, A) = \frac{\theta\left(|\omega \mp \mu_B^* H| - A\right)}{\sqrt{(\omega \mp \mu_B^* H)^2 - A^2}}, \tag{8}$$

the upper (lower) sign corresponds to the majority (minority) spin orientation, e is the electron charge, R is the "normal state" (above T_d) resistance of the junction, $0 \le \mu \le 1$ is is the relative portion of the FS sections gapped by CDWs, $\theta(x)$ denotes the Heaviside theta function,

$$D(T) = \left[\Delta^2(T) + \Sigma^2(T)\right]^{1/2} \tag{9}$$

is an "effective" gap on the "dielectrized" FS sections and it can be shown [57] that it is the BCS-Mühlschlegel function, i.e. $D(T) = \Delta_{\mathrm{BCS}}(\Sigma_0, T)$. The quantity $\Sigma_0 \equiv \frac{\pi}{\gamma} T_d$ is the magnitude of the CDW order parameter for $T = 0$ and in the absence of superconductivity, $\gamma = 1.78\ldots$ is the Euler constant. We suggested that quasiparticles originating from all FS sections make their contributions to the total current proportional to the DOS of the relevant section. That means the absence of any kind of the directional tunneling, which is possible, in principle [63–65]. Such an assumtion may be justified by the inevitable spatial averaging over CDW domains with different wave vector orientations.

The important difference between the problem in point and its counterpart appropriate to the BCS superconductivity is the emergence of the terms $J_{ib,\uparrow(\downarrow)}$. They reflect the existence of the electron-hole pairing [24, 29], originate from the interband Green's function $\mathcal{G}_{12}$, and have another structure than the remaining terms induced by conventional normal Green's functions $\mathcal{G}_{11} = \mathcal{G}_{22}$ and $\mathcal{G}_{33}$ (see discussion in Refs. [60, 62]). To a large extent $\mathcal{G}_{12}$ is analogous to the anomalous Gor'kov Green's function $\mathcal{F}$, which, however, determines not a quasiparticle but a Josephson tunnel current [66]. The appearance of the terms (6) leads to the drastic *asymmetry* of the CVC of non-symmetrical tunnel junctions involving CDWMs [62] as opposed to *symmetrical* CVC for similar non-symmetrical junctions based on conventional superconductors [67]. It should be born in mind that those current components depend on the sign of Σ, whereas the thermodynamical properties of CDW superconductors are degenerate with respect to this sign [36, 68].

When a CDW metal is normal, the expressions for the components (5) and (6) remain the same with an accuracy of $|\Sigma|$ substituted for D. At the same time, the nd components are calculated explicitly for arbitrary T:

$$J_{nd,\downarrow(\uparrow)} = \frac{(1-\mu)\,(1 \mp P)\,V}{2R}, \tag{10}$$

and the current contribution $J_{nd} = J_{nd,\downarrow} + J_{nd,\uparrow}$ from the nd FS section obeys the Ohm's law

$$J_{nd} = (1-\mu)\frac{V}{R}, \tag{11}$$

Conductivities $G_{f,s}(V)$ can be obtained by differentiating relevant Eqs. (4)-(6). At $T = 0$, the corresponding analytical expressions become

$$G_{nd,\downarrow(\uparrow)}(V) = \frac{(1-\mu)\,(1 \pm P)}{2R}\mathrm{sgn}(V)\,(eV \mp \mu_B H)\,f_{\mp}(eV, H, \Delta), \tag{12}$$

$$G_{d,\downarrow(\uparrow)}(V) = \frac{\mu\,(1 \pm P)}{2R}\mathrm{sgn}(V)\,(eV \mp \mu_B H)\,f_{\mp}(eV, H, D), \tag{13}$$

$$G_{ib,\downarrow(\uparrow)}(V) = \frac{\mu\,(1 \pm P)\,\Sigma}{2R}\mathrm{sgn}(V) f_{\mp}(eV, H, D). \tag{14}$$

Naturally, for normal CDW metals the sum of the nd terms gives the constant $\frac{(1-\mu)}{R}$.

3.3 Results and discussion

Below we show the results obtained for the dependences of the dimensionless conductance RdJ/dV of the CDWM-I-FM junction on the

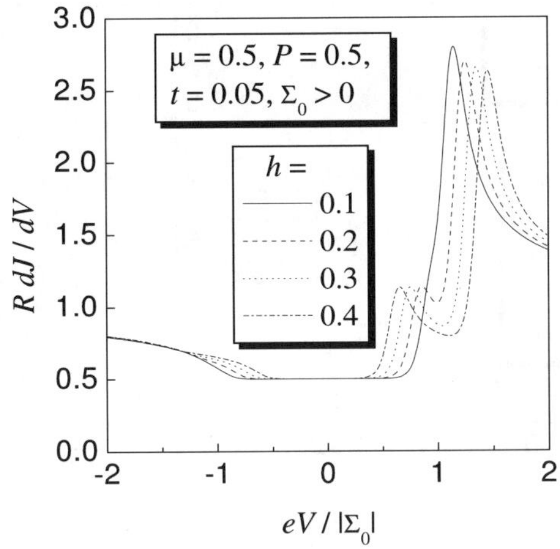

Figure 1. Dependences of differential conductances on bias voltage V across the tunnel junction made up of charge-density wave metal (CDWM) and ferromagnet (FM) for various external magnetic fileds H. See explanations in the text.

dimensionless bias voltage $eV/|\Sigma_0|$. The dimensionless parameters of the problem are the reduced external magnetic field $h = \mu_B^* H/|\Sigma_0|$, the reduced temperature $t = k_B T/|\Sigma_0|$ and the polarization P. Here μ_B^* is the effective Bohr magneton and k_B is the Boltzmann constant.

The key result of this paper is represented by Fig. 1. It is readily seen that $G(V)$ is asymmetrical, contrary to what is appropriate for superconductors [67]. Mathematically it stems from an almost total compensation between $G_d(V)$ and $G_{ib}(V)$ peculiarities at voltages of one sign and their enhancement at voltages of the other sign (for the adopted choice $\Sigma > 0$ it means negative and positive V, respectively). In the absence of the external magnetic field and spin polarization this result was obtained by us earlier [61, 62]. When H is switched on, the electronic DOS peak splits as in the case of superconductors [1, 18]. Unfortunately, the spin-splitting is noticeable only for a certain branch ($V > 0$ for the case $\Sigma > 0$; see below). Thus, although a simple algebraic procedure of Tedrow and Meservey of finding P from values of G for certain V and H, deduced for S-I-FM junctions [1, 5], seems to fail for CDWM-I-FM ones, the advantage of the set-up proposed here to detect spin-polarization-induced changes consists in the amplification of the spin-splitting effect for one CVC branch and a larger scale of Σ in comparison to Δ.

Nevertheless, CVCs are very sensitive to the value of P. Moreover, they crucially depend on the sign of Σ in CDWM. Let us first consider the case $\Sigma > 0$ [Fig. 2, panel (a)]. One can see how the phenomenon of

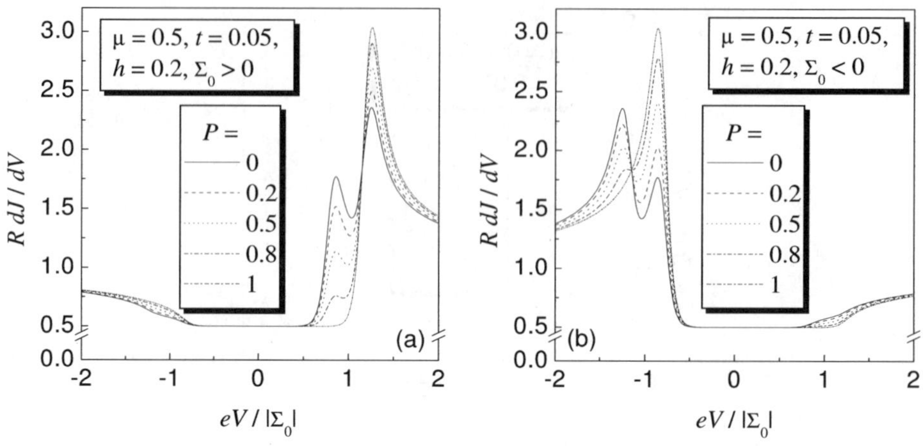

Figure 2. The same as in Figure 1 but for various FM polarizations P. Panels correspond to different sign of the dielectric order parameter Σ in CDWM. See explanations in the text.

spin-splitting reveals itself under the action of the magnetic field when a CDWM constitutes a tunnel junction with a nonmagnetic electrode ($P = 0$) and how for a ferromagnetic counter-electrode ($P \neq 0$) this picture is distorted and the minority-spin peak, which is situated closer to the zero bias, disappears with increasing P, so that for the complete polarization ($P = 1$, this limit is attainable in half-metallic ferromagnet [69–72]) only one (majority) peak retains.

For the case $\Sigma < 0$ [Fig. 2, panel (b)] the minority-spin peak also disappears with increasing P, but now it is situated farther from the zero bias than the majority one. Hence, the "modified" symmetry realtionship

$$G(-\Sigma, V) = G(\Sigma, -V) \tag{15}$$

obtained [61, 62] for junctions made up of nonferromagnetic normal and/or superconducting CDW electrodes (cf. $P = 0$ curves on both panels) is no more valid. Then different Σ signs can be distinguished by CVC measurements. It is worth to underline once more that the actual Σ sign in a specific junction might occur at random, induced by unpredictable fluctuations, since the free energy of CDW normal or superconducting metals does not depend on this sign [36, 68].

One should also bear in mind the possibility of the CVC fluctuation-induced "symmetrization" if a hypothetical small extra term in the system Hamiltonian proportional to $eV\Sigma$ exists [59]. Then the measured CVC would consist of different bias branches for $\Sigma > 0$ and $\Sigma < 0$ cases, respectively. For nonmagnetic electrodes this phenomenon, due to the

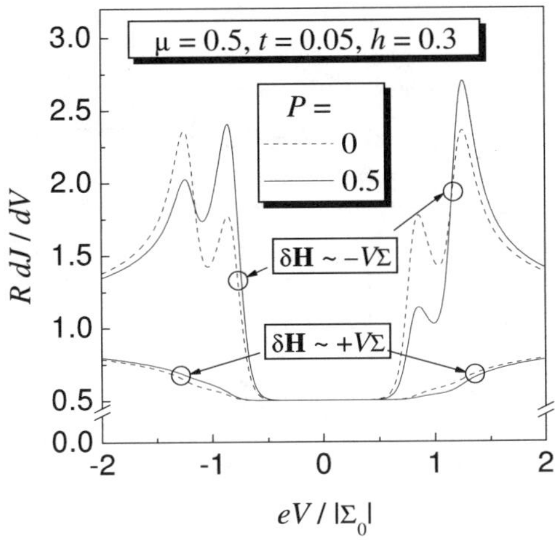

Figure 3. Hypothetical "symmetrization" effect for CDWM-FM tunnel junction. See explanations in the text.

Eq. (15), might result in a totally symmetrical CVC (see Fig. 3, dashed curves). But for $P \neq 0$ the relation (15) is not fulfiled and the nonsymmetricity of CVCs becomes unavoidable (solid curves). Unfortunately, such CVCs are possible where the peculiarities are almost unnoticable, although the CDW order parameter Σ is nonzero and may be arbitrarily large.

It is natural that all many-body features mentioned above are due to the gapped FS sections, so that when the gapping degree μ decreases, the spin-splitting and the very $G(V)$ peculiarities at $eV = |\Sigma| \pm \mu_B^* H$ are reduced and disappear, as is demonstrated in Fig. 4. The controlling parameter μ can be changed *in situ*, e. g., by application of an external pressure. Furthermore, CVCs turned out to be a sensitive probe of μ.

The smoothing effect of temperature is shown in Fig. 5. Already at a relatively small value $t = 0.2$ the Zeeman splitting becomes unobservable. However, since we can select CDW metals with T_d's of the order of $10 - 15\,\mathrm{K}$ to ensure the accesible magnetic fields $H \approx 180 - 280\,\mathrm{kOe}$, temperatures required to detect paramagnetic effects will be quite convenient from the technical point of view.

If a CDW metal becomes superconducting at $T_c < T_d$, which is not a rare case [24], two gaps Δ and D emerge on FS sections and, generally speaking, it should be four spin-splitted peaks for each voltage sign. Usually both gaps differ substantially in amplitude, as is the case, e.g., in $2H$-TaS_2, where $T_c \approx 0.65\,\mathrm{K}$ and $T_d \approx 77\,\mathrm{K}$, or in $Li_{0.9}Mo_6O_{17}$, where $T_c \approx 1.7\,\mathrm{K}$ and $T_d \approx 25\,\mathrm{K}$. The A15 compounds are the only exceptions, for which T_c and T_d are of the same order of magnitude [24, 73]. As

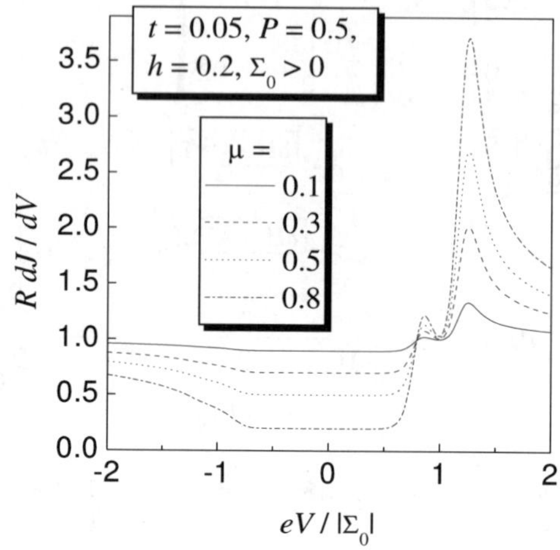

Figure 4. The same as in Figure 1 but for various degrees μ of the Fermi surface degeneracy in the CDWM.

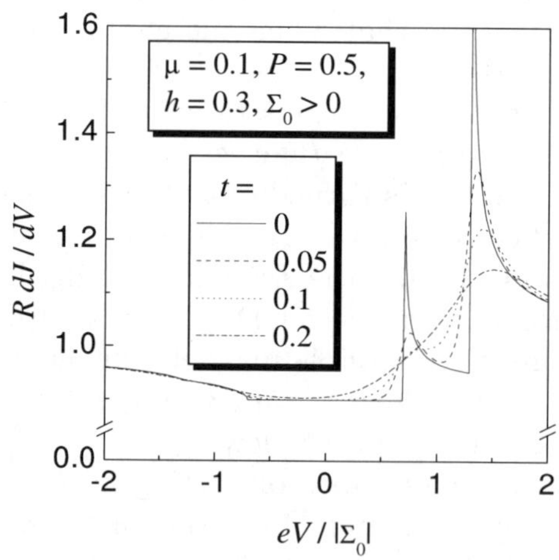

Figure 5. The same as in Figure 1 but for various temperatures T.

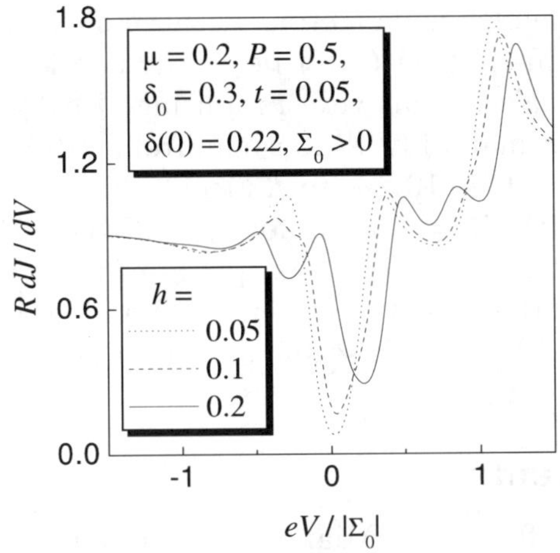

Figure 6. The same as in Figure 1 but for the tunnel junction made up of charge-density wave superconductor (CDWS) and FM. See explanations in the text.

an example, we considered an intermediate situation with the control parameter $\delta_0 \equiv \Delta_0/|\Sigma_0| = 0.3$ (see Fig. 6). Here Δ_0 is a superconducting gap for $T = 0$ in a hypothetical state where the CDW order parameter is absent. The actual $\Delta(0) = (\Delta_0 \Sigma_0^{-\mu})^{\frac{1}{1-\mu}}$ suppressed by the CDWs is smaller than Δ_0, namely, for the chosen set of parameters $\delta(0) \equiv \Delta(0)/\Sigma_0 = 0.22$. The numerical results are somewhat unexpected, since for the positive V-branch we see a paramagnetic splitting of the D-peak only (here we can distinguish merely the majority peak for Δ). At the same time, for negative biases, for which the whole D-region is almost structureless, a spin-splitting of the Δ-peak is clearly seen. This asymmetry should be observed for any $P \geq 0$ with a noticeable minority-spin contribution to the electronic DOS. Thus, the same measurements can discriminately reveal spin-dependent properties determined by both superconductivity and CDWs. At the same time, those "additional" peaks may make applicable here the Tedrow's and Meservey's procedure of determining P of ferromagnet counter-electrodes.

Of course, the type of asymmetry displayed in Fig. 6 is appropriate only to CDWs with $\Sigma > 0$. For $\Sigma < 0$ the CVC branches will have different properties with splitted D-peaks manifesting themselves for $V < 0$ and splitted Δ-peaks revealed for $V > 0$.

In conclusion, we would like to indicate several possible candidates for the CDW partner of ferromagnets in tunnel sandwiches. Organic CDW metals α-$(ET)_2MHg(SCN)_4$ (M = K, Tl, Rb) have been already mentioned. The main weak point of this materials is the presence of a heavy element Hg, which is dangerous because of the possible spin-orbit smearing of the spin-splitted $G(V)$ peaks. A two-leg ladder compound $Sr_{14-x}Ca_xCu_{24}O_{41}$ also seems very promising. Really, Ca doping alters T_d and $|\Sigma|$ over a remarkably wide range from 210 K and 130 meV, respectively, for $x = 0$ to 10 K and 130 meV for $x = 9$ [74]. The old good 2H-$NbSe_2$ with $T_c = 7.2$ K and $T_d = 33.5$ K [24] might be taken into account too. On the whole, the spread of the fruitful ideas earlier applied to superconductors [1, 18] to normal and superconducting metals partially gapped by CDWs seems useful for studying those strongly correlated objects.

Acknowledgments

A. M. G. is grateful to the Mianowski Foundation for support of his visit to Warsaw University. The research has been partly supported by the NATO grant PST.CLG.979446.

References

[1] Meservey, R. and Tedrow, P. M. (1994) Spin-polarized electron tunneling, *Phys. Rep.* **238**, 173–243.

[2] Bode, M. (2003) Spin-polarized scanning tunnelling microscopy, *Rep. Prog. Phys.* **66**, 523–582.

[3] Mazin, I. I. (1999) How to define and calculate the degree of spin polarization in ferromagnets, *Phys. Rev. Lett.* **83**, 1427–1430.

[4] Tedrow, P. M. and Meservey, R. (1971) Spin-dependent tunneling into ferromagnetic nickel, *Phys. Rev. Lett.* **26**, 192–195.

[5] Tedrow, P. M. and Meservey, R. (1973) Spin polarization of electron tunneling from films of Fe, Co, Ni, and Gd, *Phys. Rev.* **B7**, 318–326.

[6] Maki, K. and Tsuneto, T. (1964) Pauli paramagnetism and superconducting state, *Prog. Theor. Phys.* **31**, 945–956.

[7] Tsymbal, E. Y., Mryasov, O. N., and LeClair, P. R. (2003) Spin-dependent tunnelling in magnetic tunnel junctions, *J. Phys.: Condens. Matter* **15**, R109–R142.

[8] Stearns, M. B. (1977) Simple explanation of tunneling spin-polarization of Fe, Co, Ni and its alloys, *J. Magn. Magn. Materials* **5**, 167–171.

[9] Slonczewski, J. C. (1989) Conductance and exchange coupling of two ferromagnets separated by a tunneling barrier, *Phys. Rev.* **B39**, 6995–7002.

[10] Gregg, J. F., Petej, I., Jouguelet, E., and Dennis, C. (2002) Spin electronics - a review, *J. Phys.* **35**, R121–R155.

[11] Abrikosov, A. A. and Gor'kov, L. P. (1962) Spin-orbit inetraction and the Knight shift in superconductors, *Zh. Éksp. Teor. Fiz.* **42**, 1088–1096 [*Sov. Phys. JETP*].

[12] Moodera, J. S., Nassar, J., and Mathon, G. (1999) Spin-tunneling in ferromagnetic junctions, *Annu. Rev. Mater. Sci.* **29**, 381–432.

[13] Chandrasekhar, B. S. (1962) A note on the maximum critical field of high-field superconductors, *Appl. Phys. Lett.* **1**, 7–8.

[14] Clogston, A. M. (1962) Upper limit for the critical field in hard superconductors, *Phys. Rev. Lett.* **9**, 266–267.

[15] Maki, K. (1964) Pauli paramagnetism and superconducting state. II, *Prog. Theor. Phys.* **32**, 29–36.

[16] De Gennes, P. G. (1966) *Superconductivity of Metals and Alloys*, W A Benjamin, New York.

[17] Maki, K. (1969) in R. D. Parks (ed.), *Superconductivity*, Dekker, New York, Vol. 2, Chap. 18, pp. 1035–1105.

[18] Fulde, P. (1973) High field superconductivity in thin films, *Adv. Phys.* **22**, 667–719.

[19] Bulaevskii, L. N. (1975) *Usp. Fiz. Nauk* **115**, 263–300.

[20] Bulaevskii, L. N. (1975) *Usp. Fiz. Nauk* **116**, 449.

[21] Friend, R. H. and Jérome, D. (1979) *J. Phys.* **C12**, 1441–1477.

[22] Friedel, J. and Jérome, D. (1982) Organic superconductors: the $(TMTSF)_2X$ family, *Contemp. Phys.* **23**, 583–624.

[23] Wilson, J. A., Di Salvo, F. J., and Mahajan, S. (1975) Charge-density waves and superlattices in the metallic layered transition metal dichalcogenides, *Adv. Phys.* **24**, 117–201.

[24] Gabovich, A. M., Voitenko, A. I., and Ausloos, M. (2002) Charge-density waves and spin-density waves in existing superconductors: competition between Cooper pairing and Peierls or excitonic instabilities, *Phys. Rep.* **367**, 583–709.

[25] McKenzie, R. H. Is the ground state of α-(BEDT-TTF)$_2$MHg(SCN)$_4$[M=K,Rb,Tl] a charge-density wave or a spin-density wave?, cond-mat/9706235.

[26] Bjeliš, A., Zanchi, D., and Montambaux, G. Pauli and orbital effects of magnetic field on charge density waves, cond-mat/9909303.

[27] Lebed, A. G. (2003) Theory of magnetic field-induced charge-density-wave phases, *Pis'ma Zh. Éksp. Teor. Fiz.* **78**, 170–174.

[28] Jérome, D., Rice, T. M., and Kohn, W. (1967) Excitonic insulator, *Phys. Rev.* **158**, 462–475.

[29] Kopaev, Yu. V. (1975) About the interplay theory between electron and structural transformations and superconductivity, *Trudy Fiz. Inst. Akad. Nauk SSSR* **86**, 3–100.

[30] Nekovee, M. and Inglesfield, J. E. (1995) Theory of image states at magnetic interfaces, *Progr. Surf. Sci.* **50**, 149–158.

[31] Inoue, J. and Itoh, H. (2002) Tunnel magnetoresistance and interfacial electronic state, *J. Phys.* **35**, 2432–2436.

[32] Moodera, J. S. and Mathon, G. (1999) Spin polarized tunneling in ferromagnetic junctions, *J. Magn. Magn. Materials* **200**, 248–273.

[33] Zhang, X-G. and Butler, W. H. (2003) Band structure, evanescent states, and transport in spin tunnel junctions, *J. Phys.: Condens. Matter* **15**, R1603–R1639.

[34] Bilbro, G. and McMillan, W. L. (1976) Theoretical model of superconductivity and the martensitic transformation in A15 compounds, *Phys. Rev.* **B14**, 1887–1892.

[35] Halperin, B. I. and Rice, T. M. (1968) The excitonic state at the semiconductor-semimetal transition, *Solid State Phys.* **21**, 115–192.

[36] Gabovich, A. M., Gerber, A. S., and Shpigel, A. S. (1987) Thermodynamics of superconductors with charge- and spin-density waves. Δ/T_c ratio and paramagnetic limit, *Phys. Status Solidi* **B141**, 575–587.

[37] Abrikosov, A. A. (1987) *Fundamentals of the Theory of Metals*, North-Holland, Amsterdam.

[38] Bychkov, Yu. A., Gor'kov, L. P., and Dzyaloshinskii, I. E. (1966) On the possibility of the superconductivity-like phenomena in a one-dimensional system, *Zh. Éksp. Teor. Fiz.* **50**, 738–758 [(1966) *Sov. Phys. JETP* **23**, 489].

[39] Gabovich, A. M., Pashitskii, E. A., and Shpigel, A. S. (1979) About the diamagnetism of the mixed phase superconductor - excitonic insulator, *Fiz. Tverd. Tela* **21**, 463–466.

[40] Chaikin, P. M. (1985) Magnetic-field-indiced transition in quasi-two-dimensional systems, *Phys. Rev.* **B31**, 4770–4772.

[41] Bjeliš, A. and Maki, K. (1990) Magnetic-field influence on the collective properties of charge- and spin-density waves, *Phys. Rev.* **B42**, 10275–10279.

[42] Bjeliš, A. and Maki, K. (1992) Spin-density-wave and charge-density-wave phson coherence lengths in magnetic fields, *Phys. Rev.* **B45**, 12887–12892.

[43] Gor'kov, L. P. (1984) *Usp. Fiz. Nauk* **144**, 381.

[44] Maki, K. (1986) Thermodynamics of field-induced spin-density-wave states in Bechgaard salts, *Phys. Rev.* **B33**, 4826–4829.

[45] Lebed, A. G. (2002) Field-induced spin-density-wave phases in quasi-one-dimensional conductors: Theory versus experiments, *Phys. Rev. Lett.* **88**, 177001.

[46] Kwak, J. F., Schirber, J. E., Chaikin, P. M., Williams, J. M., Wang, H-H., and Chiang, L. Y. (1986) Spin-density-wave transitions in a magnetic field, *Phys. Rev. Lett.* **56**, 972–975.

[47] Audouard, A. and Askenazy, S. (1995) Spin-density wave transition and the resistivity minimum of the Bechgaard salt $(TMTSF)_2NO_3$ at high magnetic field, where TMTSF is tetramethyltetraselenafulvakene, *Phys. Rev.* **B52**, 700–703.

[48] Gweon, G-H., Denlinger, J. D., Clack, J. A., Allen, J. W., Olson, C. G., DiMasi, E., Aronson, M. C., Foran, B., and Lee, S. (1998) Direct observation of complete Fermi surface, imperfect nesting, and gap anisotropy in the high-temperature incommensurate charge-density-wave compound $SmTe_3$, *Phys. Rev. Lett.* **81**, 886–889.

[49] Williamson, S. J., Ting, C. S., and Fung, H. K. (1974) Influence of electronic lifetime on the lattice instability of V_3Si, *Phys. Rev. Lett.* **32**, 9–12.

[50] Qualls, J. S., Balicas, L., Brooks, J. S., Harrison, N., Montgomery, L. K., and Tokumoto, M. (2003) Competition between Pauli and orbital effects in a charge-density-wave system, *Phys. Rev.* **B62**, 10008–10012.

[51] Singleton, J. (2000) Studies of quasi-two-dimensional organic conductors based on BEDT-TTF using high magnetic fields, *Rep. Prog. Phys.* **63**, 1111–1207.

[52] Andres, D., Kartsovnik, M. V., Biberacher, W., Weiss, H., Balthes, E., Müller, H., and Kushch, N. (2001) Orbital effect of a magnetic field on the low-temperature state in the organic metal α-(BEDT-TTF)$_2$KHg(SCN)$_4$, *Phys. Rev.* **B64**, 161104.

[53] Harrison, N., Mielke, C. H., Christianson, A. D., Brooks, J. S., and Tokumoto, M. (2001) Field-induced dynamic diamagnetism in a charge-density-wave system, *Phys. Rev. Lett.* **86**, 1586–1589.

[54] Harrison, N. (2002) Nonequilibrium persistent currents in charge-density-wave systems, *Phys. Rev.* **B66**, 121101.

[55] Matos, M., Bonfait, G., Henriques, R. T., and Almeida, M. (1996) Modification of the magnetic-field dependence of the Peierls transition by a magnetic chain, *Phys. Rev.* **B54**, 15307–15313.

[56] Harrison, N. (1999) Destabilization of a charge-density wave by an oscillatory chemical potential, *Phys. Rev. Lett.* **83**, 1395–1398.

[57] Gabovich, A. M., Li, M. S., Szymczak, H., and Voitenko, A. I. (2003) Thermodynamics of superconductors with charge-density waves, *J. Phys.: Condens. Matter* **15**, 2745–2753.

[58] Larkin, A. I. and Ovchinnikov, Yu. N. (1966) Tunnel effect between superconductors in an alternating field, *Zh. Éksp. Teor. Fiz.* **51**, 1535–1543 [(1966) *Sov. Phys. JETP* **24**, 1035].

[59] Gabovich, A. M. and Voitenko, A. I. (1995) Tunneling spectroscopy of normal metals with charge-density or spin-density waves, *Phys. Rev.* **B52**, 7437–7447.

[60] Gabovich, A. M. and Voitenko, A. I. (1997) Nonstationary Josephson effect for superconductors with charge-density waves, *Phys. Rev.* **B55**, 1081–1099.

[61] Gabovich, A. M. and Voitenko, A. I. (1997) Josephson tunnelling involving superconductors with charge-density waves, *J. Phys.: Condens. Matter* **9**, 3901–3920.

[62] Gabovich, A. M. and Voitenko, A. I. (1997) Asymmetrical tunneling between similar metallic junctions with charge-density or spin-density waves: the case of broken symmetry, *Phys. Rev.* **B56**, 7785–7788.

[63] Dowman, J. E., MacVicar, M. L. A., and Waldram, J. R. (1969) Selection rule for tunneling from superconductors, *Phys. Rev.* **186**, 452–455.

[64] Ledvij, M. and Klemm, R. A. (1995) Dependence of the Josephson coupling of unconventional superconductors on the properties of the tunneling barrier, *Phys. Rev.* **B51**, 3269–3272.

[65] Yusof, Z., Zasadzinski, J. F., and Coffey, L. (1998) Modeling of tunneling spectroscopy in high-T_c superconductors incorporating band structure, gap symmetry, group velocity, and tunneling directionality, *Phys. Rev.* **B58**, 514–521.

[66] Kulik, I. O. and Yanson, I. K. (1970) *Josephson Effect in Superconducting Tunnel Structures*, Nauka, Moscow, in Russian.

[67] Solymar, L. (1972) *Superconductive Tunneling and Applications*, Chapman and Hall, London.

[68] Gabovich, A. M. and Shpigel, A. S. (1984) Thermodynamics of superconductors with charge- and spin-density waves, *J. Phys.* **F14**, 3031–3039.

[69] Coey, J. M. D., Viret, M., and von Molnar, S. (1999) Mixed-valence manganites, *Adv. Phys.* **48**, 167–293.

[70] Coey, J. M. D. and Venkatesan, M. (2002) Half-metallic ferromagnetism: Example of CrO2 (invited), *J. Appl. Phys.* **91**, 8345–8350.

[71] Edwards, D. M. (2002) Ferromagnetism and electron-phonon coupling in the manganites, *Adv. Phys.* **51**, 1259–1318.

[72] Irkhin, V. Yu. and Katsnelson, M. I. (1994) Half-metallic ferromagnets, *Usp. Fiz. Nauk* **164**, 705–724.

[73] Pan, V. M., Prokhorov, V. G., and Shpigel, A. S. (1984) *Metal Physics of Superconductors*, Naukova Dumka, Kiev, in Russian.

[74] Vuletić, T., Korin-Hamzić, B., Tomić, S., Gorshunov, B., Haas, P., Rõõm, T., Dressel, M., Akimitsu, J., Sasaki, T., and Nagata, T. (2003) Suppression of the charge-density-wave state in $Sr_{14}Cu_{24}O_{41}$ by calcium doping, *Phys. Rev. Lett.* **90**, 257002.

ELECTRONIC STRUCTURE OF STRONGLY CORRELATED MATERIALS: TOWARDS A FIRST PRINCIPLES SCHEME

Silke Biermann
Centre de Physique Theorique,
Ecole Polytechnique, 91128 Palaiseau, France
biermann@cpht.polytechnique.fr

Ferdi Aryasetiawan
Research Institute for Computational Sciences, AIST,
1-1-1 Umezono, Tsukuba Central 2, Ibaraki 305-8568, Japan
f-aryasetiawan@aist.go.jp

Antoine Georges
Centre de Physique Theorique,
Ecole Polytechnique, 91128 Palaiseau, France
georges@cpht.polytechnique.fr

Abstract We review a recent proposal of a first principles approach to the electronic structure of materials with strong electronic correlations. The scheme combines the GW method with dynamical mean field theory, which enables one to treat strong interaction effects. It allows for a parameter-free description of Coulomb interactions and screening, and thus avoids the conceptual problems inherent to conventional "LDA+DMFT", such as Hubbard interaction *parameters* and double counting terms. We describe the application of a simplified version of the approach to the electronic structure of nickel yielding encouraging results. Finally, open questions and further perspectives for the development of the scheme are discussed.

Keywords: Strongly correlated electron materials, first principles description of magnetism, dynamical mean field theory, GW approximation

S. Halilov (ed.), Physics of Spin in Solids: Materials, Methods and Applications, 43–65.

Introduction

The development of density functional theory in combination with rapidly increasing computing power has led to revolutionary progress in electronic structure theory over the last 40 years. Nowadays, calculations for materials with large unit cells are feasible, and applications to biological systems and complex materials of high technological importance are within reach.

The situation is less favorable, however, for so-called strongly correlated materials, where strong localization and Coulomb interaction effects cause density functional theory within the local density approximation (DFT-LDA) to fail. These are typically materials with partially filled d- or f-shells. Failures of DFT-LDA reach from missing satellite structures in the spectra (e.g. in transition metals) over qualitatively wrong descriptions of spectral properties (e.g. certain transition metal oxides) to severe qualitative errors in the calculated equilibrium lattice structures (e.g. the absence of the Ce $\alpha - \gamma$ transition in LDA or the 30% error on the volume of δ-Pu). While the former two situations may be at least partially blamed on the use of a ground state theory in the forbidden range of excited states properties, in the latter cases one faces a clear deficiency of the LDA. We are thus in the puzzling situation of being able to describe certain very complex materials, heading for a first principles description of biological systems, while not having successfully dealt with others that have much simpler structures but resist an LDA treatment due to a particular challenging electronic structure. The fact that many of these strongly correlated materials present unusual magnetic, optical or transport properties has given additional motivation to design electronic structure methods beyond the LDA.

Both, LDA+U [1–4] and LDA+DMFT [5–8] techniques are based on an Hamiltonian that explicitly corrects the LDA description by corrections stemming from local Coulomb interactions of partially localized electrons. This Hamiltonian is then solved within a static or a dynamical mean field approximation in LDA+U or LDA+DMFT respectively. In a number of magnetically or orbitally ordered insulators the LDA underestimation of the gap is successfully corrected by LDA+U (e.g. late transition metal oxides and some rare earth compounds); however its description of low energy properties is too crude to describe correlated metals, where the dynamical character of the mean field is crucial and LDA+DMFT thus more successful. Common to both approaches is the need to determine the Coulomb parameters from independent (e.g. "constrained LDA") calculations or to fit them to experiments. Neither of them thus describes long-range Coulomb interactions and the resulting screening from first principles. This has led to a recent proposal [9] of a first principles electronic structure method, dubbed "GW+DMFT" that we review in this article. Similar advances have been presented in a model context [10].

The paper is organized as follows: in section 1 we give a short overview over the parent theories (GW, DMFT, and LDA+DMFT), while in section 2 we introduce a formal way of constructing approximations by means of a free energy functional. The form of this functional defining the GW+DMFT scheme is discussed in section 3; section 4 presents different conceptual as well as technical issues related to this scheme. Finally we present results of a preliminary static implementation combining GW and DMFT, and conclude the paper with some remarks on further perspectives for the development of the GW+DMFT scheme.

4.1 The parent theories

The *GW* Approximation

Even if density functional theory is strictly only applicable to ground state properties, band dispersions of sp-electron semi-conductors and insulators have been found to be surprisingly reliable – apart from a systematic underestimation of band gaps (e.g. by $\sim 30\%$ in Si and Ge).

This underestimation of bandgaps has prompted a number of attempts at improving the LDA. Notable among these is the GW approximation (GWA), developed systematically by Hedin in the early sixties [11]. He showed that the self-energy can be formally expanded in powers of the screened interaction W, the lowest term being iGW, where G is the Green function. Due to computational difficulties, for a long time the applications of the GWA were restricted to the electron gas. With the rapid progress in computer power, applications to realistic materials eventually became possible about two decades ago. Numerous applications to semiconductors and insulators reveal that in most cases the GWA [12–14] removes a large fraction of the LDA band-gap error. Applications to alkalis show band narrowing from the LDA values and account for more than half of the LDA error (although controversy about this issue still remains [15]).

The GW approximation relies on Hedin's equations [11], which state for the self-energy that

$$\Sigma(1,2) = -i \int d3\, d4\, v(1,4) G(1,3) \frac{\delta G^{-1}(3,2)}{\delta \phi(4)} \tag{1}$$

where v is the bare Coulomb interaction, G is the Green function and ϕ is an external time-dependent probing field. We have used the shorthand notation $1 = (x_1 t_1)$. From the equation of motion of the Green function

$$G^{-1} = i\frac{\partial}{\partial t} - H_0 - \Sigma \tag{2}$$

$$H_0 = h_0 + \phi + V_H \tag{3}$$

h_0 is the kinetic energy and V_H is the Hartree potential. We then obtain

$$\begin{aligned}\frac{\delta G^{-1}(3,2)}{\delta\phi(4)} &= -\delta(3-2)\left[\delta(3-4) + \tfrac{\delta V_H(3)}{\delta\phi(4)}\right] - \tfrac{\delta\Sigma(3,2)}{\delta\phi(4)} \\ &= -\delta(3-2)\epsilon^{-1}(3,4) - \tfrac{\delta\Sigma(3,2)}{\delta\phi(4)} \end{aligned} \tag{4}$$

where ϵ^1 is the inverse dielectric matrix. The GWA is obtained by neglecting the vertex correction $\delta\Sigma/\delta\phi$, which is the last term in (4). This is just the random-phase approximation (RPA) for ϵ^{-1}. This leads to

$$\Sigma(1,2) = iG(1,2)W(1,2) \tag{5}$$

where we have defined the screened Coulomb interaction W by

$$W(1,2) = \int d3v(1,3)\epsilon^{-1}(3,2) \tag{6}$$

The RPA dielectric function is given by

$$\epsilon = 1 - vP \tag{7}$$

where

$$\begin{aligned} P(\mathbf{r},\mathbf{r}';\omega) &= -2i\int\frac{d\omega'}{2\pi}G(\mathbf{r},\mathbf{r}';\omega+\omega')G(\mathbf{r}',\mathbf{r};\omega') \\ &= 2\sum_i^{occ}\sum_j^{unocc}\psi_i(\mathbf{r})\psi_i^*(\mathbf{r}')\psi_j^*(\mathbf{r})\psi_j(\mathbf{r}') \qquad (8) \\ &\times\left\{\frac{1}{\omega-\varepsilon_j+\varepsilon_i+i\delta} - \frac{1}{\omega+\varepsilon_j-\varepsilon_i-i\delta}\right\} \qquad (9) \end{aligned}$$

with the Green function constructed from a one-particle band structure $\{\psi_i, \varepsilon_i\}$. The factor of 2 arises from the sum over spin variables. In frequency space, the self-energy in the GWA takes the form

$$\Sigma(r,r';\omega) = \frac{i}{2\pi}\int d\omega' e^{i\eta\omega'}G(r,r';\omega+\omega')W(r,r';\omega') \tag{10}$$

We have so far described the zero temperature formalism. For finite temperature we have

$$P(\mathbf{r},\mathbf{r}';i\nu_n) = \frac{2}{\beta}\sum_{\omega_k} G(\mathbf{r},\mathbf{r}';i\nu_n + i\omega_k)G(\mathbf{r}',\mathbf{r};i\omega_k) \tag{11}$$

$$\Sigma(r,r';i\omega_n) = -\frac{1}{\beta}\sum_{\nu_k} G(r,r';i\omega_n + i\nu_k)W(r,r';i\nu_k) \tag{12}$$

In the Green function language, the Fock exchange operator in the Hartree-Fock approximation (HFA) can be written as iGv. We may therefore regard the GWA as a generalization of the HFA, where the bare Coulomb interaction v is replaced by a screened interaction W. We may also think of the GWA as a mapping to a polaron problem where the electrons are coupled to some bosonic excitations (e.g., plasmons) and the parameters in this model are obtained from first-principles calculations.

The replacement of v by W is an important step in solids where screening effects are generally rather large relative to exchange, especially in metals. For example, in the electron gas, within the GWA exchange and correlation are approximately equal in magnitude, to a large extent cancelling each other, modifying the free-electron dispersion slightly. But also in molecules, accurate calculations of the excitation spectrum cannot neglect the effects of correlations or screening. The GWA is physically sound because it is qualitatively correct in some limiting cases [19].

The success of the GWA in sp materials has prompted further applications to more strongly correlated systems. For this type of materials the GWA has been found to be less successful. Application to ferromagnetic nickel [16] illustrates some of the difficulties with the GWA. Starting from the LDA band structure, a one-iteration GW calculation does reproduce the photoemission quasiparticle band structure rather well, as compared with the LDA one where the 3d band width is too large by about 1 eV. However, the too large LDA exchange splitting (0.6 eV compared with experimental value of 0.3 eV) remains essentially unchanged. Moreover, the famous 6 eV satellite, which is of course missing in the LDA, is not reproduced. These problems point to deficiencies in the GWA in describing short-range correlations since we expect that both exchange splitting and satellite structure are influenced by on-site interactions. In the case of exchange splitting, long-range screening also plays a role in reducing the HF value and the problem with the exchange splitting indicates a lack of spin-dependent interaction in the GWA: In the GWA the spin dependence only enters in G but not in W.

The GWA rather successfully improves on the LDA errors on the bandgaps of Si, Ge, GaAs, ZnSe or Diamond, but for some materials, such as MgO and InN, a significant error still remains. The reason for the discrepancy has not been understood well. One possible explanation is that the result of the one-iteration GW calculation may depend on the starting one-particle band structure, since the starting Green's function is usually constructed from the LDA Kohn-Sham orbitals and energies. For example, in the case of InN, the starting LDA band structure has no gap. This may produce a metal-like (over)screened interaction W which fails to open up a gap or yields a too small gap in the GW calculation. A similar behaviour is also found in the more extreme case of NiO, where a one-iteration GW calculation only yields a gap of about 1 eV starting from an LDA gap of 0.2 eV (the experimental gap is 4 eV) [17, 12]. This problem may be circumvented by performing a partial self-consistent calculation in which the self-energy from the previous iteration at a given energy, such as the Fermi energy of the centre of the band of interest, is used to construct a new set of one-particle orbitals. This procedure is continued to self-consistency such that the starting one-particle band structure gives zero self-energy correction [17, 12, 18]. In the case of NiO this procedure improves the band gap considerably to a self-consistent value of 5.5 eV and at the same time increases the LDA magnetic moment from 0.9 μ_B to about 1.6 μ_B much closer to the experimental value of 1.8 μ_B. A more serious problem, however, is describing the charge-transfer character of the top of the valence band. Charge-transfer insulators are characterized by the presence of occupied and unoccupied 3d bands with the oxygen 2p band in between. The gap is then formed by the oxygen 2p and unoccupied 3d bands, unlike the gap in LDA, which is formed by the 3d states. A more appropriate interpretation is to say that the highest valence state is a charge-transfer state: During photoemission a hole is created in the transition metal site but due to the strong 3d Coulomb repulsion it is energetically more favourable for the hole to hop to the oxygen site despite the cost in energy transfer. A number of experimental data, notably 2p core photoemission resonance, suggest that the charge-transfer picture is more appropriate to describe the electronic structure of transition metal oxides. The GWA, however, essentially still maintains the 3d character of the top of the valence band, as in the LDA, and misses the charge-transfer character dominated by the 2p oxygen hole. A more recent calculation using a more refined procedure of partial self-consistency has also confirmed these results [18]. The problem with the GWA appears to arise from inadequate account of short-range correlations, probably not properly treated in the random-phase approximation (RPA). As in nickel, the problem with the satellite arises again in NiO. Depending on the starting band structure, a satellite may be reproduced albeit at a too high energy. Thus there is a strong need for improving the short-range correlations in the GWA which may be achieved by using a suitable approach based on the dynamical mean-field theory described in the next section.

Dynamical Mean Field Theory

Dynamical mean field theory (DMFT) [20] has originally been developed within the context of *models* for correlated fermions on a lattice where it has proven very successful for determining the phase diagrams or for calculations of excited states properties. It is a non-perturbative method and as such appropriate for systems with any strength of the interaction. In recent years, combinations of DMFT with band structure theory, in particular Density functional theory with the local density approximation (LDA) have emerged [5, 6]. The idea is to correct for shortcomings of DFT-LDA due to strong Coulomb interactions and localization (or partial localization) phenomena that cause effects very different from a homogeneous itinerant behaviour. Such signatures of correlations are well-known in transition metal oxides or f-electron systems but are also present in several elemental transition metals.

Combinations of DFT-LDA and DMFT, so-called "LDA+DMFT" techniques have so far been applied – with remarkable success – to transition metals (Fe, Ni, Mn) and their oxides (e.g. La/$YTiO_3$, V_2O_3, Sr/$CaVO_3$, Sr_2RuO_4) as well as elemental f-electron materials (Pu, Ce) and their compounds [7]. In the most general formulation, one starts from a many-body Hamiltonian of the form

$$\begin{aligned} H &= \sum_{\{im\sigma\}} (H^{LDA}_{im,i'm'} - H^{dc}) a^{+}_{im\sigma} a_{i'm'\sigma} \\ &+ \frac{1}{2} \sum_{imm'\sigma} U^{i}_{mm'} n_{im\sigma} n_{im'-\sigma} \\ &+ \frac{1}{2} \sum_{im \neq m'\sigma} (U^{i}_{mm'} - J^{i}_{mm'}) n_{im\sigma} n_{im'\sigma}, \end{aligned} \tag{13}$$

where H^{LDA} is the effective Kohn-Sham-Hamiltonian derived from a self-consistent DFT-LDA calculation. This one-particle Hamiltonian is then corrected by Hubbard terms for direct and exchange interactions for the "correlated" orbitals, e.g. d or f orbitals. In order to avoid double counting of the Coulomb interactions for these orbitals, a correction term H^{dc} is subtracted from the original LDA Hamiltonian. The resulting Hamiltonian (13) is then treated within dynamical mean field theory by assuming that the many-body self-energy associated with the Hubbard interaction terms can be calculated from a multi-band impurity model.

This general scheme can be simplified in specific cases, e.g. in systems with a separation of the correlated bands from the "uncorrelated" ones, an effective model of the correlated bands can be constructed; symmetries of the crystal structure can be used to reduce the number of components of the self-energy etc.

In this way, the application of DMFT to *real solids* crucially relies on an appropriate definition of the local screened Coulomb interaction U (and Hund's rule coupling J). DMFT then assumes that local quantities such as for example the local Green's function or self-energy of the solid can be calculated from a local impurity model, that is one can find a *dynamical mean field* $\mathcal{G}_0$ such that the Green's function calculated from the effective action

$$\begin{aligned} S &= \int_0^\beta d\tau \sum_{m\sigma} c^\dagger_{m\sigma}(\tau) \mathcal{G}^{-1}_{0mm'\sigma}(\tau-\tau') c_{m'\sigma}(\tau') \\ &+ \frac{1}{2}\int_0^\beta d\tau \sum_{mm'\sigma} U_{mm'} n_{m\sigma}(\tau) n_{m'-\sigma}(\tau) \\ &+ \frac{1}{2}\int_0^\beta d\tau \sum_{m\neq m'\sigma} (U_{mm'} - J_{mm'}) n_{m\sigma}(\tau) n_{m'\sigma}(\tau) \end{aligned} \tag{14}$$

coincides with the local Green's function of the solid. This is in analogy to the *representability conjecture* of DMFT in the model context, where one assumes that e.g. the local Green's function of a Hubbard model can be represented as the Green's function of an appropriate impurity model with the same U parameter. In the case of a lattice with infinite coordination number it is trivially seen that this conjecture is correct, since DMFT yields the exact solution in this case. Also, the model context is simpler from a conceptual point of view, since the Hubbard U is given from the outset. For a real solid the situation is somewhat more complicated, since in the construction of the impurity model *long-range* Coulomb interactions are mimicked by *local* Hubbard parameters.[1] The notion of locality is also needed for the resolution of the model within DMFT, which approximates the full self-energy of the model by a local quantity. Applications of DMFT to electronic structure calculations (e.g. the LDA+DMFT method) are therefore always defined within a specific basis set using localized basis functions. Within an LMTO [21] implementation for example locality can naturally be defined as referring to the same muffin tin sphere. This amounts to defining matrix elements

$G_{L\mathbf{R},L'\mathbf{R}'}(i\omega)$ of the full Green's function

$$G(\mathbf{r},\mathbf{r}',i\omega) = \sum_{LL'\mathbf{R}\mathbf{R}'} \chi^*_{L\mathbf{R}}(\mathbf{r}) G_{L\mathbf{R},L'\mathbf{R}'}(i\omega)\chi_{L'\mathbf{R}'}(\mathbf{r}') \tag{15}$$

and assuming that its local, that is "on-sphere" part equals the Green's function of the local impurity model (13). Here $\mathbf{R},\mathbf{R}'$ denote the coordinates of the centres of the muffin tin spheres, while $\mathbf{r}$, $\mathbf{r}'$ can take any values. The index $L = (n, l, m)$ regroups all radial and angular quantum numbers. The dynamical mean field $\mathcal{G}_0$ in (13) has to be determined in such a way that the Green's function $G_{impurityL,L'}$ of the impurity model Eq.(13) coincides with $G_{L\mathbf{R},L'\mathbf{R}}(i\omega)$ if the impurity model self-energy is used as an estimate for the true self-energy of the solid. This self-consistency condition reads

$$G_{impurity}(i\omega_n) = \sum_k (i\omega_n + \mu - H_o(k) - \Sigma(i\omega_n))^{-1}$$

where Σ, H_0 and G are matrices in orbital and spin space, and $i\omega + \mu$ is a matrix proportional to the unit matrix in that space.

Together with (13) this defines the DMFT equations that have to be solved self-consistently. Note that the main approximation of DMFT is hidden in the self-consistency condition where the local self-energy has been promoted to the full lattice self-energy.

The representability assumption can actually be extended to other quantities of a solid than its local Green's function and self-energy. In "extended DMFT" [22–25] e.g. a two particle correlation function is calculated and can then be used in order to represent the local screened Coulomb interaction W of the solid. This is the starting point of the "GW+DMFT" scheme described in section 6.

Despite the huge progress made in the understanding of the electronic structure of correlated materials thanks to such LDA+DMFT schemes, certain conceptual problems remain open: These are related to the choice of the Hubbard interaction parameters and to the double counting corrections. An a priori choice of which orbitals are treated as correlated and which orbitals are left uncorrelated has to be made, and the values of U and J have to be fixed. Attempts of calculating these parameters from constrained LDA techniques are appealing in the sense that one can avoid introducing external parameters to the theory, but suffer from the conceptual drawback in that screening is taken into account in a static manner only [26]. Finally, the double counting terms are necessarily ill defined due to the impossibility to single out in the LDA treatment contributions to the interactions stemming from specific orbitals. These drawbacks of LDA+DMFT provide a strong motivation to attempt the construction of an electronic structure method for correlated materials beyond combinations of LDA and DMFT.

4.2 The Ψ-functional

As noted in [27, 28], the free energy of a solid can be viewed as a functional $\Gamma[G, W]$ of the Green's function G and the screened Coulomb interaction W. The functional Γ can trivially be split into a Hartree part Γ_H and a many body correction Ψ, which contains all corrections beyond the Hartree approximation : $\Gamma = \Gamma_H + \Psi$. The latter is the sum of all skeleton diagrams that are irreducible with respect to both, one-electron propagator and interaction lines. $\Psi[G, W]$ has the following properties:

$$\frac{\delta\Psi}{\delta G} = \Sigma^{xc}$$
$$\frac{\delta\Psi}{\delta W} = P. \tag{16}$$

We present in the following a different derivation than the one given in [27].

We start from the Hamiltonian describing interacting electrons in an external (crystal) potential v_{ext} :

$$H = H_0 + \frac{1}{2}\sum_{i\neq j} V_{ee}(\mathbf{r}_i - \mathbf{r}_j) \tag{17}$$

with

$$H_0 = -\frac{1}{2}\sum_i \left(\nabla_i^2 + v_{ext}\right). \tag{18}$$

The electron-electron interaction $V_{ee}(\mathbf{r}_i - \mathbf{r}_j)$ will later on be assumed to be a Coulomb potential. With the action

$$\begin{aligned} S &= -\int d\tau \Psi^\dagger(\delta_\tau - H_0 - V_{Hartree})\Psi \\ &+ \frac{1}{2}\int d\tau \ :n\left(\mathbf{r}\right): V\left(\mathbf{r}-\mathbf{r}'\right): n\left(\mathbf{r}'\right): \end{aligned} \tag{19}$$

the partition function of this system reads :

$$Z = \int \mathbf{D}\Psi\mathbf{D}\Psi^\dagger exp(-S) \tag{20}$$

Here the double dots denote normal ordering (i.e. Hartree terms have been included in the first term in 18). We now do a Hubbard-Stratonovich transform, decoupling Coulomb interactions beyond Hartree by a continuous field ϕ, introduce a coupling constant α for later purposes ($\alpha = 1$ corresponds to the original problem) and add source terms coupling to the density fluctuations $\Psi^\dagger\Psi$ and the density of the Hubbard-Stratonovich field respectively . The free energy of the system is now a functional of the source fields Σ and P :

$$F[\Sigma, P] = \ln \int \mathbf{D}\Psi\mathbf{D}\Psi^\dagger\mathbf{D}\phi exp(-S[\Sigma, P]) \tag{21}$$

with

$$\begin{aligned} S[\Sigma, P] &= -\int d\tau \Psi^\dagger G_{Hartree}\Psi + \frac{1}{2}\int d\tau\ \phi V^{-1}\phi \\ &- i\alpha \int d\tau \phi(\Psi^\dagger\Psi - n) + \int d\tau \Sigma\Psi^\dagger\Psi + \frac{1}{2}\int d\tau P\phi\phi \end{aligned} \tag{22}$$

If in analogy to the usual fermionic Green's function $G = -\langle \mathbf{T}\Psi\Psi^\dagger \rangle$ we define the propagator $W = \langle \mathbf{T}\phi\phi \rangle$ of the Hubbard-Stratonovich field ϕ, our specific choice of the coupling of the sources Σ and P leads to

$$\frac{\delta F}{\delta \Sigma} = -G \tag{23}$$

and

$$\frac{\delta F}{\delta P} = \frac{1}{2} W \tag{24}$$

Performing Legendre transformations with respect to G and $W/2$ we obtain the free energy as a functional of both, the fermionic and bosonic propagators G and W

$$\Gamma[G, W] = F + tr(\Sigma G) - \frac{1}{2} tr(PW) \tag{25}$$

We note that W can be related to the charge-charge response function $\chi(\mathbf{r}, \mathbf{r}'; \tau - \tau') \equiv \langle T_\tau[\hat{\rho}(\mathbf{r}, \tau) - n(\mathbf{r})][\hat{\rho}(\mathbf{r}', \tau') - n(\mathbf{r}')] \rangle$:

$$\begin{aligned} W(\mathbf{r}, \mathbf{r}'; i\omega_n) &= V(\mathbf{r} - \mathbf{r}') \\ &- \int d\mathbf{r}_1 d\mathbf{r}_2 V(\mathbf{r} - \mathbf{r}_1)\chi(\mathbf{r}_1, \mathbf{r}_2; i\omega_n)V(\mathbf{r}_2 - \mathbf{r}') \end{aligned} \tag{26}$$

This property follows directly from the above functional integral representation and justifies the identification of W with the screened Coulomb interaction in the solid. Taking advantage of the coupling constant α introduced above we find that Γ is naturally split into two parts

$$\Gamma_{\alpha=1}[G, W] = \Gamma_{\alpha=0} + \Psi[G, W] \tag{27}$$

where the first is just the Hartree free energy

$$\begin{aligned}\Gamma_{\alpha=0}(G,W) &= Tr\ln G - Tr[(G_H^{-1}-G^{-1})G] \\ &- \frac{1}{2}Tr\ln W + \frac{1}{2}Tr[(v^{-1}-W^{-1})W] \end{aligned} \tag{28}$$

(with $G_H^{-1} = i\omega_n + \mu + \nabla^2/2 - V_H$ denoting the Hartree Green's function and V_H the Hartree potential), while the second one

$$\Psi = \int_0^1 d\alpha \frac{\delta\Gamma}{\delta\alpha} \tag{29}$$

contains all non-trivial many-body effects. Needless to mention, this correlation functional Ψ cannot be calculated exactly, and approximations have to be found.

The GW approximation consists in retaining the first order term in α only, thus approximating the Ψ-functional by

$$\Psi[G,W] = -\frac{1}{2}Tr(GWG). \tag{30}$$

We then find trivially

$$\Sigma = \frac{\delta\Psi}{\delta G} = -GW \tag{31}$$

$$P = \frac{\delta\Psi}{\delta W} = GG. \tag{32}$$

4.3 GW+DMFT

Inspired by the description of screening within the GW approximation and the great successes of DMFT in the description of strongly correlated materials, the "GW+DMFT" method [9] is constructed to retain the advantages of both, GW and DMFT, without the problems associated to them separately. In the GW+DMFT scheme, the Ψ-functional is constructed from two elements: its local part is supposed to be calculated from an impurity model as in DMFT, while its non-local part is given by the non-local part of the GW Ψ-functional:

$$\Psi = \Psi_{GW}^{\mathrm{non-loc}}[G^{\mathbf{RR}'}, W^{\mathbf{RR}'}] + \Psi_{imp}[G^{\mathbf{RR}}, W^{\mathbf{RR}}] \tag{33}$$

Since the strong correlations present in materials with partially localized d- or f-electrons are expected to be much stronger in their local components than in their non-local ones the non-local physics is assumed to be well described by a perturbative treatment as in GW, while local physics is described within an impurity model in the DMFT sense.

More explicitly, the non-local part of the GW+DMFT Ψ-functional is given by

$$\Psi_{GW}^{\text{non-loc}}[G^{\mathbf{RR}'}, W^{\mathbf{RR}'}] = \Psi_{GW}[G^{\mathbf{RR}'}, W^{\mathbf{RR}'}] - \Psi_{GW}^{\text{loc}}[G^{\mathbf{RR}'}, W^{\mathbf{RR}'}] \tag{34}$$

while the local part is taken to be an impurity model Ψ functional. Following (extended) DMFT, this onsite part of the functional is generated from a local *quantum impurity problem* (defined on a single atomic site). The expression for its free energy functional $\Gamma_{imp}[G_{imp}, W_{imp}]$ is analogous to (27) with $\mathcal{G}$ replacing G_H and $\mathcal{U}$ replacing V :

$$\begin{aligned} \Gamma_{imp}[G_{imp}, W_{imp}] &= Tr \ln G_{imp} - Tr[(\mathcal{G}^{-1} - G_{imp}^{-1})G_{imp}] \\ &- \frac{1}{2} Tr \ln W_{imp} + \frac{1}{2} Tr[(\mathcal{U}^{-1} - W_{imp}^{-1})W_{imp}] \\ &+ \Psi_{imp}[G_{imp}, W_{imp}] \end{aligned} \tag{35}$$

The impurity quantities G_{imp}, W_{imp} can thus be calculated from the effective action:

$$\begin{aligned} S &= \int d\tau d\tau' \left[- \sum c_L^\dagger(\tau) \mathcal{G}_{LL'}^{-1}(\tau - \tau') c_{L'}(\tau') \right. \\ &+ \left. \frac{1}{2} \sum : c_{L_1}^\dagger(\tau) c_{L_2}(\tau) : \mathcal{U}_{L_1 L_2 L_3 L_4}(\tau - \tau') : c_{L_3}^\dagger(\tau') c_{L_4}(\tau') : \right] \end{aligned} \tag{36}$$

where the sums run over all orbital indices L. In this expression, $c_L^\dagger$ is a creation operator associated with orbital L on a given sphere, and the double dots denote normal ordering (taking care of Hartree terms).

The construction (33) of the Ψ-functional is the only ad hoc assumption in the GW+DMFT approach. The explicit form of the GW+DMFT equations follows then directly from the functional relations between the free energy, the Green's function, the screened Coulomb interaction etc. Taking derivatives of (33) as in (15) it is seen that the complete self-energy and polarization operators read:

$$\begin{aligned} \Sigma^{xc}(\mathbf{k}, i\omega_n)_{LL'} &= \Sigma_{GW}^{xc}(\mathbf{k}, i\omega_n)_{LL'} \\ &- \sum_{\mathbf{k}} \Sigma_{GW}^{xc}(\mathbf{k}, i\omega_n)_{LL'} + [\Sigma_{imp}^{xc}(i\omega_n)]_{LL'} \end{aligned} \tag{37}$$

$$\begin{aligned} P(\mathbf{q}, i\nu_n)_{\alpha\beta} &= P^{GW}(\mathbf{q}, i\nu_n)_{\alpha\beta} \\ &- \sum_{\mathbf{q}} P^{GW}(\mathbf{q}, i\nu_n)_{\alpha\beta} + P^{imp}(i\nu_n)_{\alpha\beta} \end{aligned} \tag{38}$$

The meaning of (37) is transparent: the off-site part of the self-energy is taken from the GW approximation, whereas the onsite part is calculated to all orders from the dynamical impurity model. This treatment thus goes beyond usual E-DMFT, where the lattice self-energy and polarization are just taken to be their impurity counterparts. The second term in (37) substracts the onsite component of the GW self-energy thus avoiding double counting. At self-consistency this term can be rewritten as:

$$\sum_{\mathbf{k}} \Sigma^{xc}_{GW}(\tau)_{LL'} = - \sum_{L_1 L_1'} W^{imp}_{LL_1 L' L_1'}(\tau) G_{L_1' L_1}(\tau) \qquad (39)$$

so that it precisely substracts the contribution of the GW diagram to the impurity self-energy. Similar considerations apply to the polarization operator.

From a technical point of view, we note that while one-particle quantities such as the self-energy or Green's function are represented in the localized basis labeled by L, two-particle quantities such as P or W are expanded in a two-particle basis, labeled by Greek indices in (38). We will discuss this point more in detail below.

We now outline the iterative loop which determines $\mathcal{G}$ and $\mathcal{U}$ self-consistently (and, eventually, the full self-energy and polarization operator):

- The impurity problem (36) is solved, for a given choice of $\mathcal{G}_{LL'}$ and $\mathcal{U}_{\alpha\beta}$: the "impurity" Green's function

  $$G^{LL'}_{imp} \equiv -\langle T_\tau c_L(\tau) c^+_{L'}(\tau')\rangle_S \qquad (40)$$

  is calculated, together with the impurity self-energy

  $$\Sigma^{xc}_{imp} \equiv \delta\Psi_{imp}/\delta G_{imp} = \mathcal{G}^{-1} - G^{-1}_{imp}. \qquad (41)$$

  The two-particle correlation function

  $$\chi_{L_1 L_2 L_3 L_4} = \langle : c^\dagger_{L_1}(\tau) c_{L_2}(\tau) :: c^\dagger_{L_3}(\tau') c_{L_4}(\tau') :\rangle_S \qquad (42)$$

  must also be evaluated.

- The impurity effective interaction is constructed as follows:

  $$W^{\alpha\beta}_{imp} = \mathcal{U}_{\alpha\beta} - \sum_{L_1 \cdots L_4} \sum_{\gamma\delta} \mathcal{U}_{\alpha\gamma} O^{\gamma}_{L_1 L_2} \chi_{L_1 L_2 L_3 L_4} [O^{\delta}_{L_3 L_4}]^* \mathcal{U}_{\delta\beta} \qquad (43)$$

  where $O^{\alpha}_{L_1 L_2} \equiv \langle \phi_{L_1} \phi_{L_2} | B^{\alpha}\rangle$ is the overlap matrix between two-particle states and products of one-particle basis functions. The polarization operator of the impurity problem is then obtained as:

  $$P_{imp} \equiv -2\delta\Psi_{imp}/\delta W_{imp} = \mathcal{U}^{-1} - W^{-1}_{imp}, \qquad (44)$$

where all matrix inversions are performed in the two-particle basis (see the discussion at the end of this section).

- From Eqs. (37) and (38) the full $\mathbf{k}$-dependent Green's function $G(\mathbf{k}, i\omega_n)$ and effective interaction $W(\mathbf{q}, i\nu_n)$ can be constructed. The self-consistency condition is obtained, as in the usual DMFT context, by requiring that the onsite components of these quantities coincide with G_{imp} and W_{imp}. In practice, this is done by computing the onsite quantities

$$G_{loc}(i\omega_n) = \sum_{\mathbf{k}} [{G_H}^{-1}(\mathbf{k}, i\omega_n) - \Sigma^{xc}(\mathbf{k}, i\omega_n)]^{-1} \quad (45)$$

$$W_{loc}(i\nu_n) = \sum_{\mathbf{q}} [V_{\mathbf{q}}^{-1} - P(\mathbf{q}, i\nu_n)]^{-1} \quad (46)$$

and using them to update the Weiss dynamical mean field $\mathcal{G}$ and the impurity model interaction $\mathcal{U}$ according to:

$$\mathcal{G}^{-1} = G_{loc}^{-1} + \Sigma_{imp} \quad (47)$$

$$\mathcal{U}^{-1} = W_{loc}^{-1} + P_{imp} \quad (48)$$

This cycle (which is summarized in Fig.1) is iterated until self-consistency for $\mathcal{G}$ and $\mathcal{U}$ is obtained (as well as on G, W, Σ^{xc} and P). Eventually, self-consistency over the local electronic density can also be implemented, (in a similar way as in LDA+DMFT [29, 30]) by recalculating $\rho(\vec{r})$ from the Green's function at the end of the convergence cycle above, and constructing an updated Hartree potential. This new density is used as an input of a new GW calculation, and convergence over this external loop must be reached.

We stress that in this scheme the Hubbard interaction $\mathcal{U}$ is no longer an external parameter but is determined self-consistently. The appearance of the two functions $\mathcal{U}$ and W_{loc} might appear puzzling at first sight, but has a clear physical interpretation: W_{loc} is the fully screened Coulomb interaction, while $\mathcal{U}$ is the Coulomb interaction screened by all electronic degrees of freedom that are not explicitly included in the effective action. So for example onsite screening is included in W_{loc} but not in $\mathcal{U}$. From Eq. (48) it is seen that further screening of $\mathcal{U}$ by the onsite polarization precisely leads to W_{loc}.

In the following we discuss some important issues for the implementation of the proposed scheme, related to the choice of the basis functions for one-particle and two-particle quantities. In practice the self-energy is expanded in some basis set $\{\phi_L\}$ localized in a site. The polarization function on the other hand is expanded in a set of two-particle basis functions $\{\phi_L \phi_{L'}\}$ (product basis) since the polarization

$$\begin{array}{cc}
\text{Impurity model :} & \\
\mathcal{G}(\tau),\mathcal{U}(\tau) & \\
G_{imp} \equiv -\langle T_\tau \, cc^\dagger \rangle_S \;\rightarrow & \Sigma^{xc}_{imp} = \mathcal{G}^{-1} - G^{-1}_{imp} \\
W_{imp} = \mathcal{U} - \mathcal{U}\chi\mathcal{U} & P_{imp} = \mathcal{U}^{-1} - W^{-1}_{imp} \\
\uparrow & \downarrow \\
\text{Update} & \text{Combine :} \\
\mathcal{G}^{-1} = G^{-1}_{loc} + \Sigma_{imp} & \Sigma^{xc} = \Sigma^{xc}_{imp} + \Sigma^{xc,nonlocal}_{GW} \\
\mathcal{U}^{-1} = W^{-1}_{loc} + P_{imp} & P = P_{imp} + P^{nonlocal}_{GW} \\
\uparrow & \downarrow \\
\end{array}$$

$$\begin{array}{c}
\text{Self} - \text{consistency} \\
G_{loc} = \sum_{\mathbf{k}} [G_H{}^{-1} - \Sigma^{xc}]^{-1} \\
W_{loc} = \sum_{\mathbf{q}} [V_{\mathbf{q}}^{-1} - P]^{-1}
\end{array}$$

Figure 1. Schematic representation of the inner (DMFT) self-consistency cycle of the GW+DMFT scheme, consisting of the construction of the impurity model effective action (36) from the Weiss field $\mathcal{G}$ and the impurity Coulomb interaction $\mathcal{U}$, the solution of the impurity model leading to an estimate for the impurity self-energy Σ_{imp} and the polarization P_{imp} and the self-consistency condition where local quantities of the solid are calculated and then used to update the impurity model. Full self-consistency of the whole scheme requires an additional outer cycle updating the GW Hartree Potential corresponding to the obtained electronic density (cf text).

corresponds to a two-particle propagator. For example, when using the linear muffin-tin orbital (LMTO) band-structure method, the product basis consists of products of LMTO's. These product functions are generally linearly dependent and a new set of optimized product basis (OPB) [12] is constructed by forming linear combinations of product functions, eliminating the linear dependencies. We denote the OPB set by $B_\alpha = \sum_{LL'} \phi_L \phi_{L'} c^\alpha_{LL'}$. To summarize, one-particle quantities like G and Σ are expanded in $\{\phi_L\}$ whereas two-particle quantities such as P and W are expanded in the OPB set $\{B_\alpha\}$. It is important to note that the number of $\{B_\alpha\}$ is generally smaller than the number of $\{\phi_L \phi_{L'}\}$ so that quantities expressed in $\{B_\alpha\}$ can be expressed in $\{\phi_L \phi_{L'}\}$, but not vice versa. E.g. matrix elements in products of LMTOs can be obtained from those in the $\{B_\alpha\}$ basis via the transformation

$$W^{\mathbf{RR}'}_{L_1L_2L_3L_4} \equiv \langle \phi^{\mathbf{R}}_{L_1} \phi^{\mathbf{R}}_{L_2} | W | \phi^{\mathbf{R}'}_{L_3} \phi^{\mathbf{R}'}_{L_4} \rangle = \sum_{\alpha\beta} O^\alpha_{L_1L_2} W^{\mathbf{RR}'}_{\alpha\beta} O^{\beta*}_{L_3L_4} \tag{49}$$

with the overlap matrix $O^\alpha_{L_1L_2} \equiv \langle \phi_{L_1} \phi_{L_2} | B^\alpha \rangle$, but the knowledge of the matrix elements $W_{L_1L_2L_3L_4}$ alone does not allow to go back to the $W_{\alpha\beta}$.

4.4 Challenges and open questions

Global self-consistency

As has been pointed out, the above GW+DMFT scheme involves self-consistency requirements at two levels: for a given density, that is Hartree potential, the dynamical mean field loop detailed in section 4.1.0 must be iterated until $\mathcal{G}$ and $\mathcal{U}$ are self-consistently determined. This results in a solution for G, W, Σ^{xc} and P corresponding to this given Hartree potential. Then, a new estimate for the density must be calculated from G, leading to a new estimate for the Hartree potential, which is reinserted into (45). This external loop is iterated until self-consistency over the local electronic density is reached.

The external loop is analogous to the one performed when GW calculations are done self-consistently. From calculations on the homogeneous electron gas [31] it is known, that self-consistency at this level worsens the description of spectra (and in particular washes out satellite structures) when vertex corrections are neglected. Since the combined GW+DMFT scheme, however, includes the local vertex to all orders and only neglects non-local vertex corrections we expect the situation to be more favorable in this case. Test calculations to validate this hypothesis are an important subject for future studies. First steps in this direction have been undertaken in [32].

The notion of locality : the choice of the basis set

The construction of the GW+DMFT functional Ψ_{GW} crucially relies on the notion of local and nonlocal contributions. In a solid these notions can only be defined by introducing a basis set of localized functions centered on the atomic positions. Local components are then defined to be those functions the arguments of which refer to the same atomic lattice site. We stress that in this way the concept of locality is *not* a pointwise (δ-like) one. It merely means that local quantities have an expansion (15) within their atomic sphere the form of which is determined by the basis functions used. This feature is shared between LDA+DMFT and the GW+DMFT scheme, and approaches that could directly work in the continuum are so far not in sight. If however, the basis set dependence induced by this concept is of practical importance remains to be tested.

Within LDA+DMFT the choice of the basis functions enters at two stages : First, the definition of the onsite U and the quality of the approximation consisting in the neglect of offsite interactions depends on the degree of localization of the basis functions. Second, the DMFT approximation promoting a local quantity to the full self-energy of the solid is the better justified the more localized the chosen basis functions are. However, in the spirit of obtaining an accurate low-energy description one might – depending on the material under consideration – in some cases be led to work in a Wannier function basis incorporating weak hybridisation effects of more extended states with the localized ones. This then leads to slightly more extended basis functions, and one has to compromise between maximally localized orbitals and an efficient description of low-energy bands.

In GW+DMFT, some non-local corrections to the self-energy are included, so that the DMFT approximation should be less severe. More importantly, this scheme could (via the $\mathcal{U}$ self-consistency requirement) automatically adapt to more or less localized basis functions by choosing itself the appropriate $\mathcal{U}$. Therefore, the basis set dependence is likely to be much weaker in GW+DMFT than in LDA+DMFT.

Separation of correlated and uncorrelated orbitals

As mentioned in section 1 the construction of the LDA+DMFT Hamiltonian requires an *a priori* choice of which orbitals are treated as correlated or uncorrelated orbitals. Since in GW+DMFT the Hubbard interactions are determined self-consistently it might be perceivable not to perform such a separation at the outset, but to rely on the self-consistency cycle to find small values for the interaction between itinerant (e.g. s or p) orbitals. Even if this issue would probably not play a role for practical calculations, it would be satisfactory from a conceptual point of view to be able to treat all orbitals on an equal footing. It is therefore also an important question for future studies.

Dynamical impurity models

Dynamical impurity models are hard to solve, since standard techniques used for the solution of static impurity models within usual DMFT, such as the Hirsch-Fye QMC algorithm or approximate techniques such as the iterative perturbation theory are not applicable. First attempts have been made in [10] and [40] using different auxiliary field QMC schemes and in [39] using an approximate "slave-rotor" scheme. These techniques, however, have so far only been applied in the context of model systems, and their implementation in a multi-band realistic calculation is at present still a challenging project. In the following section we therefore present a simplified static implementation of the GW+DMFT scheme.

4.5 Static implementation

Here, we demonstrate the feasibility and potential of the approach within a simplified implementation, which we apply to the electronic structure of Nickel. The main simplifications made are: (i) The DMFT local treatment is applied only to the *d*-orbitals, (ii) the GW calculation is done only once, in the form [12]: $\Sigma_{GW}^{xc} = G_{LDA} \cdot W[G_{LDA}]$, from which the non-local part of the self-energy is obtained, (iii) we replace the dynamical impurity problem by its static limit, solving the impurity model (36) for a frequency-independent $\mathcal{U} = \mathcal{U}(\omega = 0)$. Instead of the Hartree Hamiltonian we start from a one-electron Hamiltonian in the form: $H_{LDA} - V_{xc,\sigma}^{nonlocal} - \frac{1}{2}\mathbf{Tr}\Sigma_{\sigma}^{imp}(0)$. The non-local part of this Hamiltonian coincides with that of the Hartree Hamiltonian while its local part is derived from LDA, with a double-counting correction of the form proposed in [33] in the DMFT context. With this choice the self-consistency condition (45) reads:

$$\begin{aligned} G_{loc}^{\sigma}(i\omega_n) &= \sum_{\mathbf{k}} [{G_H}^{-1}(\mathbf{k}, i\omega_n) - (\Sigma_{GW}^{xc})_{non-loc} \\ &- (\Sigma_{imp,\sigma} - \frac{1}{2}\mathbf{Tr}_{\sigma}\Sigma_{imp,\sigma}(0) + V_{xc}^{loc})\,]^{-1} \end{aligned} \tag{50}$$

We have performed finite temperature GW and LDA+DMFT calculations (within the LMTO-ASA[21] with 29 **k**-points in the irreducible Brillouin zone) for ferromagnetic nickel (lattice constant 6.654 a.u.), using 4s4p3d4f states, at the Matsubara frequencies $i\omega_n$ corresponding to

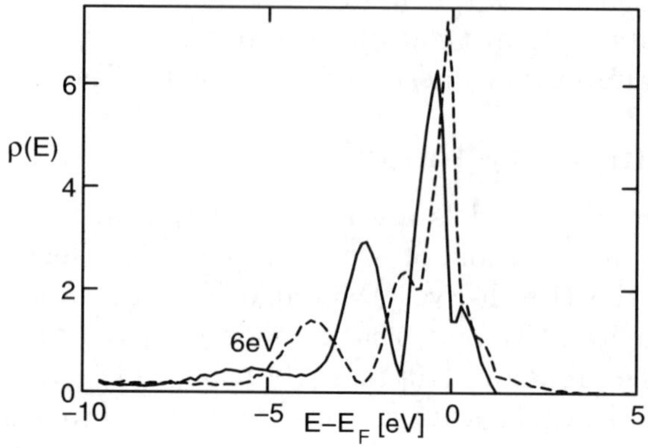

Figure 2. Partial density of states of d-orbitals of nickel (solid [dashed] lines give the majority [minority] spin contribution) as obtained from the combination of GW and DMFT (see text). For comparison with LDA and LDA+DMFT results see [33], for experimental spectra see [34].

$T = 630K$, just below the Curie temperature. The resulting self-energies are inserted into Eq. (50), which is then used to calculate a new Weiss field according to (47). The Green's function $G^{\sigma}_{loc}(\tau)$ is recalculated from the impurity effective action by QMC and analytically continued using the Maximum Entropy algorithm. The resulting spectral function is plotted in Fig.(2). Comparison with the LDA+DMFT results in [33] shows that the good description of the satellite structure, exchange splitting and band narrowing is indeed retained within the (simplified) GW+DMFT scheme.

We have also calculated the quasiparticle band structure, from the poles of (50), after linearization of $\Sigma(\mathbf{k}, i\omega_n)$ around the Fermi level [2]. Fig. (3) shows a comparison of GW+DMFT with the LDA and experimental band structure. It is seen that GW+DMFT correctly yields the bandwidth reduction compared to the (too large) LDA value and renormalizes the bands in a (**k**-dependent) manner.

We now discuss further the simplifications made in our implementation. Because of the static approximation (iii), we could not implement self-consistency on W_{loc} (Eq. (46)). We chose the value of $\mathcal{U}(\omega = 0)$ ($\simeq 3.2eV$) by calculating the correlation function χ and ensuring that Eq. (43) is fulfilled at $\omega = 0$, given the GW value for $W_{loc}(\omega = 0)$ ($\simeq 2.2eV$ for Nickel [38]). This procedure emphasizes the low-frequency, screened value, of the effective interaction. Obviously, the resulting impurity self-energy Σ_{imp} is then much smaller than the local component of the GW self-energy (or than V_{xc}^{loc}), especially at high frequencies. It

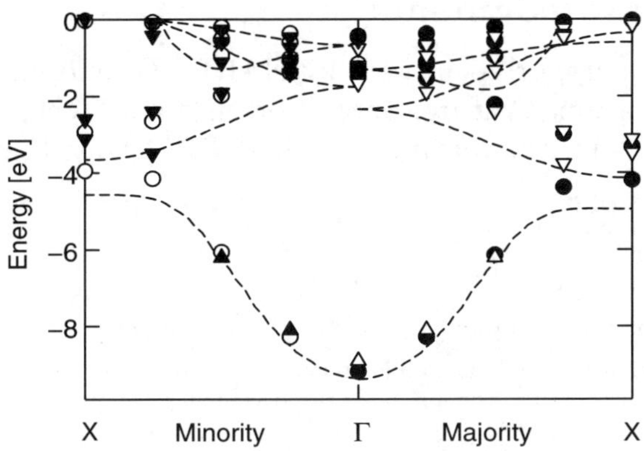

Figure 3. Band structure of Ni (minority and majority spin) as obtained from the linearization procedure of the GW+DMFT self-energy described in the text (dots) in comparison to the LDA band structure (dashed lines) and the experimental data of [35] (triangles down) and [34] (triangles up).

is thus essential to choose the second term in (37) to be the onsite component of the GW self-energy rather than the r.h.s of Eq. (39). For the same reason, we included V_{xc}^{loc} in Eq.(50) (or, said differently, we implemented a mixed scheme which starts from the LDA Hamiltonian for the local part, and thus still involves a double-counting correction). We expect that these limitations can be overcome in a self-consistent implementation with a frequency-dependent $\mathcal{U}(\omega)$ (hence fulfilling Eq. (39)). In practice, it might be sufficient to replace the local part of the GW self-energy by Σ_{imp} for correlated orbitals only. Alternatively, a downfolding procedure could be used.

4.6 Perspectives

In conclusion, we have reviewed a recent proposal of an *ab initio* dynamical mean field approach for calculating the electronic structure of strongly correlated materials, which combines GW and DMFT. The scheme aims at avoiding the conceptual problems inherent to "LDA+DMFT" methods, such as double counting corrections and the use of Hubbard parameters assigned to correlated orbitals. A full practical implementation of the GW+DMFT scheme is a major goal for future research, which requires further work on impurity models with frequency-dependent interaction parameters [41, 42, 10] as well as studies of various possible self-consistency schemes.

4.7 Acknowledgments

This work was supported in part by NAREGI Nanoscience Project, Ministry of Education, Culture, Sports, Science and Technology, Japan and by a grant of supercomputing time at IDRIS Orsay, France (project number 031393).

Notes

1. For a discussion of the appropriateness of local Hubbard parameters see [26].

2. Note however that this linearization is no longer meaningful at energies far away from the Fermi level. We therefore use the unrenormalized value for the quasi-particle residue for the s-band ($Z_s = 1$).

References

[1] V. I. Anisimov, J. Zaanen, and O. K. Andersen, Phys. Rev. B **44**, 943 (1991)

[2] V. I. Anisimov, I. V. Solovyev, M. A. Korotin, M. T. Czyzyk, and G. A. Sawatzky, Phys. Rev. B **48**, 16929 (1993)

[3] A. I. Lichtenstein, J. Zaanen, and V. I. Anisimov, Phys. Rev. B **52**, R5467 (1995)

[4] For a review, see V. I. Anisimov, F. Aryasetiawan, and A. I. Lichtenstein, J. Phys.: Condens. Matter **9**, 767 (1997)

[5] V. I. Anisimov et al., J. Phys.: Condens. Matter **9**, 7359 (1997)

[6] A. I. Lichtenstein and M. I. Katsnelson, Phys. Rev. B **57**, 6884 (1998).

[7] For reviews, see *Strong Coulomb correlations in electronic structure calculations*, edited by V. I. Anisimov, Advances in Condensed Material Science (Gordon and Breach, New York, 2001)

[8] For related ideas, see : G. Kotliar and S. Savrasov in *New Theoretical Approaches to Strongly Correlated Systems*, Ed. by A. M. Tsvelik (2001) Kluwer Acad. Publ. (and the updated version: cond-mat/0208241)

[9] S. Biermann, F. Aryasetiawan, and A. Georges, Phys. Rev. Lett. **90**, 086402 (2003)

[10] P. Sun and G. Kotliar, Phys. Rev. B **66**, 085120 (2002)

[11] L. Hedin, Phys. Rev. **139**, A796 (1965); L. Hedin and S. Lundqvist, *Solid State Physics* vol. 23, eds. H. Ehrenreich, F. Seitz, and D. Turnbull (Academic, New York, 1969)

[12] F. Aryasetiawan and O. Gunnarsson, Rep. Prog. Phys. **61**, 237 (1998)

[13] W. G. Aulbur, L. Jönsson, and J. W. Wilkins, Solid State Physics **54**, 1 (2000)

[14] G. Onida, L. Reining, A. Rubio, Rev. Mod. Phys. **74**, 601 (2002).

[15] W. Ku, A. G. Eguiluz, and E. W. Plummer, Phys. Rev. Lett. **85**, 2410 (2000); H. Yasuhara, S. Yoshinaga, and M. Higuchi, *ibid.* **85**, 2411 (2000)

[16] F. Aryasetiawan, Phys. Rev. B **46**, 13051 (1992)

[17] F. Aryasetiawan and O. Gunnarsson, Phys. Rev. Lett. **74**, 3221 (1995)

[18] S. V. Faleev, M. van Schilfgaarde, and T. Kotani, unpublished

[19] L. Hedin, Int. J. Quantum Chem. **54**, 445 (1995)

[20] For reviews, see A. Georges, G. Kotliar, W. Krauth, and M. J. Rozenberg, Rev. Mod. Phys. **68**, 13 (1996); T. Pruschke *et al*, Adv. Phys. **44**, 187 (1995)

[21] O. K. Andersen, Phys. Rev. B **12**, 3060 (1975); O. K. Andersen, T. Saha-Dasgupta, S. Erzhov, Bul. Mater. Sci. **26**, 19 (2003)

[22] Q. Si and J. L. Smith, Phys. Rev. Lett. **77**, 3391 (1996)

[23] G. Kotliar and H. Kajueter (unpublished)

[24] H. Kajueter, Ph.D. thesis, Rutgers University, 1996

[25] A. M. Sengupta and A. Georges, Phys. Rev. B **52**, 10295 (1995)

[26] F. Aryasetiawan, M. Imada, A. Georges, G. Kotliar, S. Biermann, and A. I. Lichtenstein, submitted to Phys. Rev. B.

[27] C.-O. Almbladh, U. von Barth and R. van Leeuwen, Int. J. Mod. Phys. B **13**, 535 (1999)

[28] R. Chitra and G. Kotliar, Phys. Rev. B **63**, 115110 (2001)

[29] S. Savrasov and G. Kotliar, cond-mat/0106308

[30] S. Savrasov, G. Kotliar and E. Abrahams, Nature (London) **410**, 793 (2000)

[31] B. Holm and U. von Barth, Phys. Rev. B **57**, 2108 (1998)

[32] P. Sun and G. Kotliar, cond-mat/0312303

[33] A. I. Lichtenstein, M. I. Katsnelson and G. Kotliar, Phys. Rev. Lett. **87**, 067205 (2001)

[34] H. Mårtensson and P. O. Nilsson, Phys. Rev. B **30**, 3047 (1984)

[35] J. Bünemann *et al*, Europhys. Lett. **61**, 667 (2003)

[36] Y. Motome and G. Kotliar, Phys. Rev. B **62**, 12800 (2000)

[37] J. K. Freericks, M. Jarrell and D. J. Scalapino, Phys. Rev. B **48**, 6302 (1993)

[38] M. Springer and F. Aryasetiawan, Phys. Rev. B **57**, 4364 (1998)

[39] S. Florens, PhD thesis, Paris 2003 ; S. Florens, A. Georges, L. Demedici, unpublished.

[40] A. Rubtsov, unpublished.

[41] Y. Motome, G. Kotliar, Phys. Rev. B **62**, 12800 (2000).

[42] J. K. Freericks, M. Jarrell, D. J. Scalapino, Phys. Rev. B **48**, 6302 (1993).

SPIN-DENSITY WAVE AND SHORT-RANGE OSCILLATIONS IN PHOTOEMISSION FROM FILMS OF CR METAL

S.L. Molodtsov
Institut fr Festkrperphysik, Technische Universitt Dresden,
D-01062 Dresden, Germany
molodtso@physik.phy.tu-dresden.de

Abstract

The origin of both the long- and the short-range magnetic oscillations in films of Cr metal is studied with photoemission (PE). The experimental data are analyzed on the basis of results of the electronic structure calculations performed within the local spin-density approximation (LSDA) - layer Korringa-Kohn-Rostoker (LKKR) approach and the density functional theory (DFT) using a screened-KKR Green's function method. It is shown that the incommensurate spin-density wave (SDW) can be monitored and important parameters of SDW-related interactions, such as coupling strength and energy of collective magnetic excitations, can be determined from the dispersion of the renormalized electronic bands close to the Fermi energy. The used approach can be applied to a large variety of other SDW systems including magnetic multilayer structures highly relevant for technological applications. The short-range PE intensity modulations at the Fermi energy are related to the quantum-well states (QWS), which were for the first time observed in <100> directions in Cr(100) layers. Possible contributions of the QWS into the short-range and the long-range magnetic coupling between marginal layers in Fe/Cr/Fe systems were discussed.

Keywords: Spin-density wave, quantum-well states, photoemission, chromium

In this contribution we report on mostly intriguing parts of the electronic structure of thin films of Cr metal. These parts are an incommensurate spin-density wave (SDW) state and short-range electronic density of states (DOS) oscillations observed upon variation of thicknesses of Cr layers.

The electronic properties of Cr are of high importance both for fundamental and more applied reasons. Bulk Cr is an almost unique ma-

S. Halilov (ed.), Physics of Spin in Solids: Materials, Methods and Applications, 67–84.

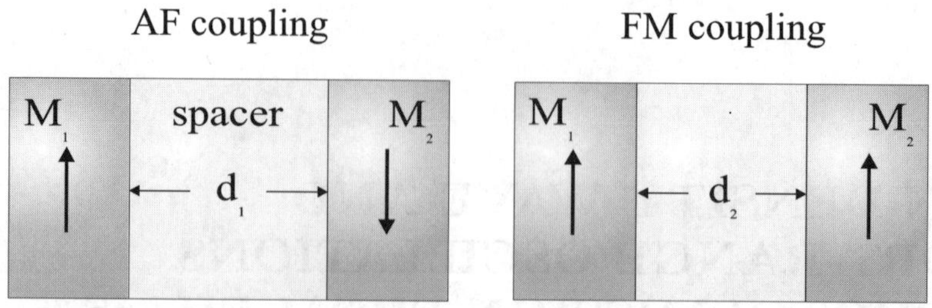

Figure 1. AF and FM couplung between magnetic moments of Fe layers (M_1 and M_2) depending on thickness of Cr spacer (d_1 and d_2).

terial revealing itinerant antiferromagnetism with the spin-density wave ground state at room temperature [1]. At the Nel temperature $T_N = 311$ K chromium exhibits a transition from a paramagnetic (*bcc* lattice) to an antiferromagnetic (AF) order (*sc* of CsCl type) that is modulated by the incommensurate SDW along the <100> directions [1–3]. Thereby the periods of the AF arrangement and the SDW oscillations amount to 2 and $\sim$ 21 monolayers (ML) of Cr, respectively. It is widely accepted that the AF order (often referred to as the commensurate SDW) is caused by a nesting of the Fermi surface (FS) sheets around the Γ and the H points of the Brillouin zone (BZ) of *bcc* Cr, while the nature of the incommensurate SDW is still the subject of extensive debates [4–7]. The SDW in Cr is accompanied by a strain wave and a charge-density wave (CDW) with half the period of the SDW [8] as well as by a series of collective excitations including spin waves (magnons) and phonons [1]. Electron interactions particularly with the magnetic excitations lead to renormalization of the electronic structure of the ground state. Although a number of attempts was made to study the renormalization of the electronic bands in some Cr systems [1, 9] the subject requires further investigations.

A detailed understanding of the SDW and the SDW-related phenomena in Cr is of primary interest, since the above short- and long-range magnetic modulations give strong reason to use Cr as spacers in magnetic multilayer structures providing giant magnetoresistance, spin-valve effect and applications in magnetic sensor technology [10]. One of the mostly investigated up to now system Fe/Cr/Fe(100) shows that the ferromagnetic (FM) or AF type of coupling between Fe layers (Fig. 1) varies with thickness of Cr spacer following the short period ($D_S = 2$ ML), whereas the strength of the coupling changes with the long period [$D_L \sim 11$ ML, about half a wavelength of the incommensurate SDW, Fig. 2] of oscillations [11–14]. Thereby, mostly accepted point

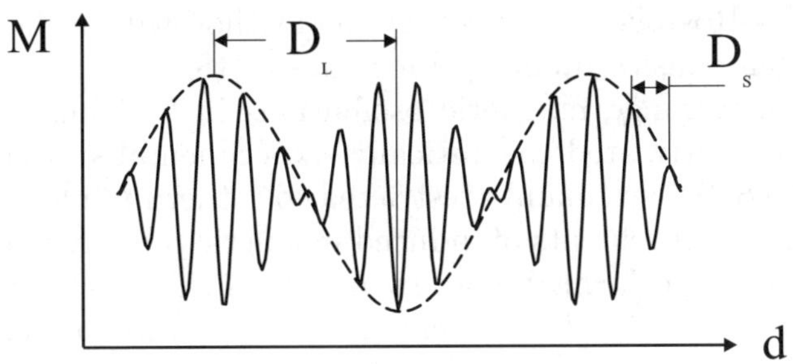

Figure 2. Short- and long-range oscillations of the magnetic coupling.

of view considers the short range of oscillations to be caused by the AF coupling between neighboring (100) ferromagnetic Cr monolayers in the spacer. Although other mechanisms like quantum-well state (QWS) or Ruderman-Kittel-Kasuya-Yosida (RKKY) [4, 7, 15] coupling are still under discussion.

More controversial is the situation around the incommensurate long-range oscillations. In their study Schilfgaarde and Harrison [4] suggested that these modulations stem from aliasing of the short-range oscillations due to a slight mismatch between the nesting vector spanning the FS sheets around the Γ and the H points of the BZ and the period of the reciprocal lattice in the <100> directions. This model, however, is not supported by the experimental results obtained by scanning electron microscopy with polarization-analysis [13]. On the other hand, it was shown that the long-range oscillations can be caused by nesting conditions characterized by smaller spanning vectors found at other sheets of the FS [6, 7].

The description of the magnetic oscillations in thin films is complicated by the fact that boundary conditions at the interfaces have to be properly considered. While nodes of the SDW are expected at Cr/Mo(100) junctions [16], interfaces with Fe(100) marginal layers reveal antinodes of the Cr magnetic moments. In the density-functional theory study of Fe/Cr/Fe(100) by Niklasson *et al.* [17] mainly AF order was found for Cr spacers with thicknesses < 10 ML. For thicker layers, various branches of sometimes coexisting SDWs, which differ from each other by the number of nodes m, were calculated. Upon increase of the Cr thickness, each m branch is abruptly substituted by a $(m+2)$ branch giving rise to phase slips of the short-range oscillations [13], which, however, may also be correlated with peculiarities of the bulk nesting conditions. The SDW order in Fe/Cr/Fe was also treated by means of the

Korringa-Kohn-Rostoker Green's function method within the framework of the local spin-density functional formalism [18].

While the long-range magnetic oscillations in Cr films seems to be relatively well investigated theoretically, experimental studies of the incommensurate SDW are mainly restricted to rather indirect information obtained from measurements of induced magnetic properties of marginal layers, which were performed, e.g., by means of the magneto-optic Kerr effect [14], spin-polarized electron-energy loss spectroscopy [12] and scanning electron microscopy with polarization analysis [13]. So far, no systematic photoemission (PE) studies of the magnetic oscillations in Cr films of different thicknesses except Refs. [7, 19, 20] were reported. On the other hand, particularly PE provides mostly direct insight into the structure of the occupied electron states allowing better understanding of the discussed phenomena.

Here we present results of PE studies of chromium films in both regimes: incommensurate SDW for thick films [21] and AF coupling for thin films, where incommensurate order is suppressed [22].

5.1 Incommensurate spin-density wave

Experimental

We studied the incommensurate SDW phenomena in Cr systems by an angle-resolved PE of epitaxial Cr films (10 to 100 ML) grown on W(110). The measurements were performed with a SCIENTA 200 electron-energy analyzer using monochromatized light from a He lamp ($h\nu = 40.8$ eV). The overall-system energy resolution including thermal broadening was set to 130 meV full width at half maximum (FWHM), the angular resolution to 0.4°. All experiments were carried out at room temperature well below T_N at the surface of Cr(110) [19, 23]. The base pressure in the experimental set up was 6×10^{-9} Pa. Films of Cr were prepared *in situ* on a W(110) crystal by deposition from a molybdenum crucible heated by electron beam. Various thicknesses of Cr were used in order to follow the transition from the thick films characterized by the SDW state (47 and 100 ML) to the thin film (10 ML), where the incommensurate SDW is not present anymore [17]. "As deposited" samples were annealed to 900°C in order to ensure well-ordered films under consideration. In all cases the grown epitaxial films revealed sharp low-energy electron diffraction patterns.

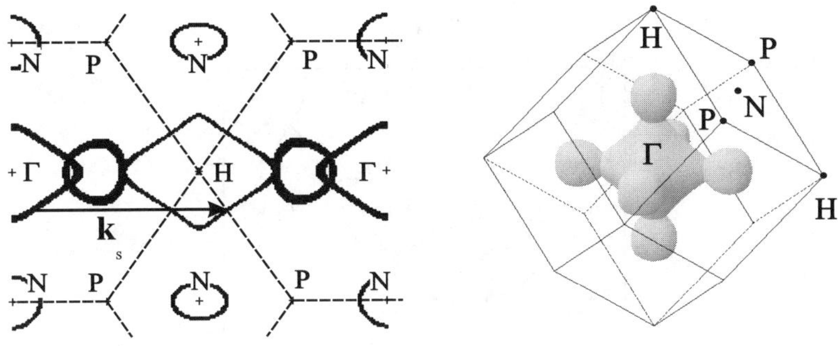

Figure 3. Cut of the calculated bulk FS of *bcc* Cr within the (110) plane (left). FS jack around the Γ point (right).

Electronic structure calculations

The electronic structure of the Cr(110) semi-infinite crystal was calculated within a local spin-density approximation (LSDA) - layer Korringa-Kohn-Rostoker (LKKR) multiple scattering approach. This method uses the Green's function technique and allows the calculation of systems with a broken translational symmetry along one direction [24]. The calculated layer resolved spectral density of states are related to the layer Green's function simply as $D(\mathbf{k}_{\|}, E) = -\mathrm{Tr}\,\mathrm{Im}\,G(\mathbf{k}_{\|}, E)/\pi$.

Experimental results and discussion

Fig. 3 (left) demonstrates a calculated cut of the bulk FS of *bcc* Cr within the (110) plane. The contours of the FS around the Γ and the H points of the bulk BZ look almost identical: They are rhomb-like and are connected by the spanning vector $\mathbf{k}_s$. Therefore, they are expected to be strongly affected by the magnetic ordering. Also the three-dimensional FS jacks at the Γ and the H points [the Γ-point jack is shown in Fig. 3 (right)] have similar shapes. In a first approximation they can be obtained from each other by a parallel transfer defined by the $\mathbf{k}_s$ vector. Therefore, everywhere in the region of these jacks one would expect the energy gap and the band renormalization related to the SDW state.

We have performed experiments along the $\overline{\Gamma} - \overline{S}$ direction in the surface BZ of Cr by varying the polar electron-emission angle. In this way the part of the FS around the Γ point, where the bumps at the corners [see Fig. 3 (right)] do not distort the measurements, was sampled. Assuming free-electron like final states, the measurements were carried out along the path in the bulk BZ as shown in Fig. 4, where for simplicity the FS calculated for *bcc* Cr is presented. The BZ for AF phase can be obtained from the *bcc* BZ by folding. The path crosses the FS sheet in

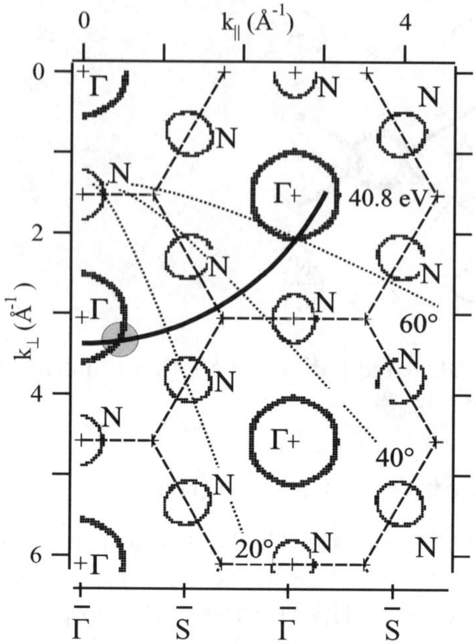

Figure 4. Path in the Brillouin zone sampled with hν = 40.8 eV (thick line). $\mathbf{k}_{||}$ and $\mathbf{k}_\perp$ denote parallel and perpendicular to the (110) plane components of the wave vector, respectively.

the region of interest marked by a shaded circle in the figure. Note that the sampling path in this region reveals a nearly zero slop and, hence, is characterized by only slightly changing $\mathbf{k}_\perp$ coordinate. Therefore, the situation we have is very similar to the case of photoemission studies of quasi-two dimensional (2D) systems (e.g., high temperature superconductors or surface states), where 2D slices of the electronic structure in the reciprocal space are considered.

Fig. 5 presents the results of the LKKR spectral DOS calculations for the bulk and the surface layer of AF *sc* Cr without the incommensurate SDW modulation. Similar to other theoretical data [3, 19, 25] the AF energy gap Δ of about 390 meV is obtained in the $\mathbf{k}_{||}$ region marked in Fig. 4. The calculated gap is predominantly located in the region of the unoccupied electron states. The corresponding "as measured" data close to the energy-gap region taken for the 100-ML thick Cr layer are shown in a gray-scale plot in Fig. 6(a). The measured band B follows the behavior of the high DOS at the boundary of a bulk state continuum. First, it approaches the Fermi energy (E_F). At $\mathbf{k}_{||} > 0.55\ ^{-1}$ it turns back toward higher binding energies (BEs). The PE intensity of the

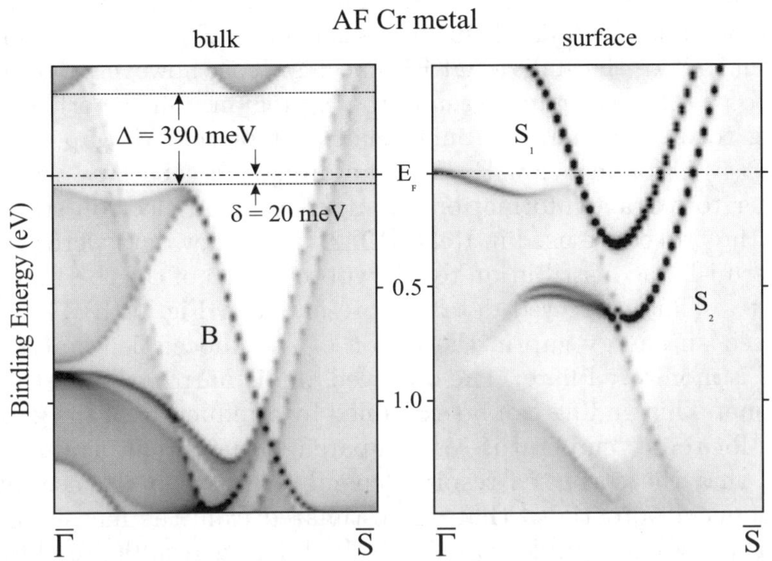

Figure 5. Spectral LKKR-DOS. $\mathbf{k}_{\|}$ along the $\overline{\Gamma} - \overline{S}$ direction. The darker the color, the more intense the DOS.

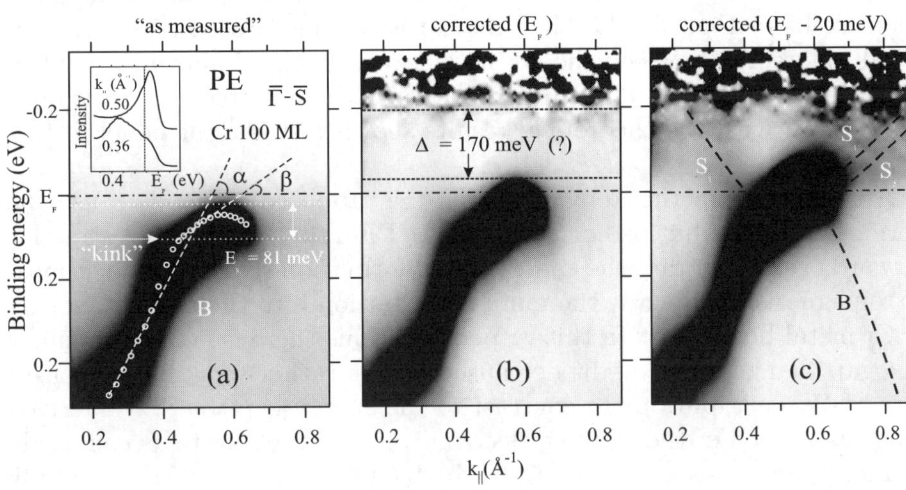

Figure 6. Logarithmic PE signal. Dark areas represent higher intensity. (a) "As measured" data. Energy of the band B maximum for each PE spectrum are marked with white circles. Inset: two spectra taken below ($\mathbf{k}_{\|} = 0.36\ ^{-1}$) and above ($0.50\ ^{-1}$) the kink position. Vertical line is the kink energy. (b) Data corrected by the Fermi-Dirac distribution with μ equal to the measured E_F. In (c) $\mu - E_F$ was selected to be -20 meV.

band in this $\mathbf{k}_{\|}$ region is almost negligible, a cross-section effect, which is well known for folded bands in solids [26]. From Fig. 6(a) and further careful analysis of the individual PE spectra, it is, however, by far not clear as to whether the gap is seen in the E_F region. On the other hand, according to our calculations only small part of the AF gap ($\delta \sim 20$ meV, Fig. 5) is found below E_F and can directly be observed with PE.

In order to extract information about unoccupied electron states we followed the procedure used in Refs. [19, 27]. The raw data were divided by the Fermi-Dirac distribution to allow observation of the thermally excited states. The corrected data are presented in Fig. 6(b). It seems that indeed an energy gap in the region of the unoccupied states of $\sim$ 170 meV is monitored here. The observed gap is narrower than the calculated one. This finding can be explained by a reduction of the gap size expected for room temperature as compared to the ground state. Note, however, that the corrected results depend strongly on the used value of the chemical potential μ that was estimated from the measurements of E_F for a metallic sample. Qualitatively different results are obtained shifting μ by only 20 meV toward higher or lower BEs, variations, which are much smaller than the energy resolution of the experiments. An increase of the chemical potential leads to a considerably larger value of the derived energy gap. Even more drastic changes are observed upon decrease of μ [Fig. 6(c)]. The AF gap is not monitored anymore. Instead, the region between the Fermi energy and 0.2 eV above E_F is filled with electron states. The surface origin of these states is seen from a comparison with the theoretical results shown in the right panel in Fig. 5.

As follows from the above analysis the proposed method to use the Fermi-Dirac distribution correction of the PE data [19, 27] is not straightforward even to extract the energy-gap information related with the antiferromagnetism, which is the main contribution into the magnetic order in Cr metal below T_N. In this respect, it seems there is no way to follow fine gap changes that might be caused by the rather weak incommensurate SDW contribution. In contrast to the energy gap the SDW-derived renormalization of the shallow electronic bands in Cr systems can easily be monitored. As seen in Fig. 5, the high DOS at the boundary of a bulk state continuum reveals smooth monotonic dispersion when going from the $\overline{\Gamma}$ point toward the gap at E_F. In difference to that our experimental band shows a pronounced "kink" at $\mathbf{k}_{\|} \sim 0.45\ ^{-1}$. The observed kink can be simulated by a superposition of two contributions: the calculated single-particle LSDA data for small $\mathbf{k}_{\|}$ and a renormalized data for $\mathbf{k}_{\|} > 0.45\ ^{-1}$. The PE spectra of band B in the region of the kink [inset in Fig. 6(a)] have an asymmetric lineshape with two structures: a main peak and a shoulder, which exchange their spectral weight crossing the kink. The main peak that is quite broad at high BE becomes much sharper close to E_F. All above evidences that the signals measured in the dispersion region before and after the kink have different nature.

This kind of behavior of bands close to E_F is a well known phenomenon for correlated systems [28]. There is a pole in the real part of the self-energy Σ of the material at the energy of a collective excitation. The strength of interaction is described via a coupling constant λ that is defined as $\lambda = -\partial Re\Sigma/\partial E|_{E_F}$. The energy dependence of Σ can be inverted into a **k** dependence. Thereby (i) the self-energy pole is transformed into the kink in the band dispersion and (ii) the coupling constant is rewritten as $\lambda = [(\partial E^{LSDA}/\partial \mathbf{k})/(\partial E^{ren}/\partial \mathbf{k})|_{E_F} - 1]$ depending on the ratio of the group velocities determined by the LSDA $[E^{LSDA}(\mathbf{k})]$ and the renormalized $[E^{ren}(\mathbf{k})]$ bands. As a result both λ and the energy of quasiparticle excitation can be derived from the analysis of the band dispersion in the vicinity of the kink. Energy gaps that can appear to stabilize the excited state may complicate the situation. In this case the quasiparticle energy is related not to E_F, but to the bottom of the corresponding gap.

The kink energy relative to the bottom of the calculated Fermi-energy gap is estimated to be (81±7) meV for the 100-ML film [Fig. 6(a)] that is of the order of the expected energy for the magnon excitations in Cr metal [1, 29]. Interaction with phonons accompanying the strain-wave or CDW state can be ruled out by the reason of much lower energy of the phonon excitations [30]. To obtain λ the ratio of the group velocities for the LSDA and the renormalized bands was substituted by the ratio of the tangents of two angles α and β between the directions of the corresponding band dispersion and the $\mathbf{k}_{\|}$ axis [see Fig. 6(a)]. The directions of band dispersion were least-square approximated by straight lines through the energy positions of the main peak of each individual spectrum that are shown by white circles in the figure. By this procedure a value $\lambda = 1.41 \pm 0.09$ was obtained pointing to a moderate strong quasiparticle interaction.

The dispersion of band B in the region of the kink was studied for Cr films of different thicknesses (Fig. 7). In all cases the films were selected to be thick enough to reveal the bulk and the surface (S_1 and S_2) features of the electronic structure of the AF *sc* Cr (Fig. 5). In this way both surface and bulk electronic properties of the bare antiferromagnetic phase of Cr metal were excluded from the consideration as possible reasons for the thickness dependent behavior of the kink, which is discussed in the following. It is of high importance to underline that no kink is observed for the 10-ML film, where the incommensurate

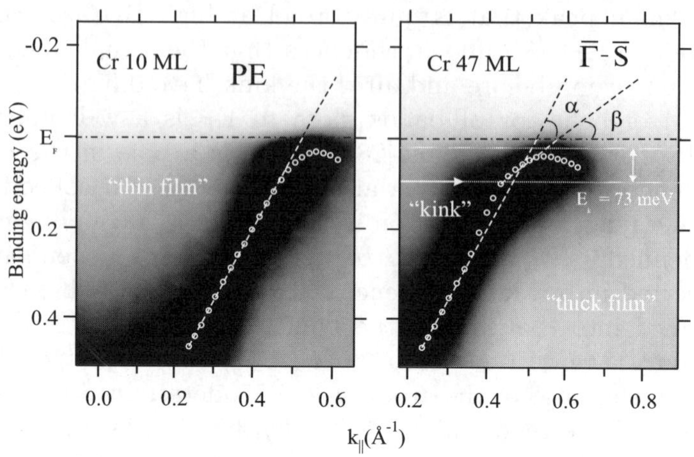

Figure 7. PE signal in the region of the kink.

SDW is suppressed [17]. Apart from the theoretical results this fact is considered as a further strong evidence for the SDW in thicker Cr films and the possibility to monitor this state in the dispersion of the renormalized bands. The disappearance of the kink allows one to expel from the analysis dimensionality effects like quantum well states, which are expected to be mostly pronounced particularly for thin films. A decrease of the film thickness from 100 to 47 ML does not result in the removal of the kink, although it causes a slight drop of its energy to (73±7) meV. Assuming a linear energy dispersion of the spin-wave excitation [1] this may be assigned to the increase of the bulk period of the related SDW due to growing influence of the boundary conditions. Also λ decreases slightly with the thickness reaching the value 1.30 ± 0.09 for the 47-ML thick film. As reported in Ref. [13] there are two phase slips in the coupling between the marginal layers in the range from 47- to 100-ML thick Cr spacers in Fe/Cr/Fe(100) systems. According to Niklasson *et al.* [17] each slip originates in a jump from one branch of the SDW to another. Therefore, the observed change of λ may be understood by slightly different electron interaction with magnons associated with individual SDW branches, which are expected to be also present in Cr/W(110).

To summarize this part of our study we have shown that angle-resolved PE can successfully be used to monitor the SDW in thin films of Cr. A valuable information about quasiparticle magnetic interactions was obtained by the analysis of the dispersion of the renormalized electronic bands in the vicinity of E_F. The used approach can be applied to a large variety of other SDW systems including magnetic multilayer structures highly relevant for technological applications.

5.2 Short-range oscillations

Experimental

The regime of the short-range oscillations was studied with thin films of chromium (thickness up to about 10 ML), where the long-range modulation phenomena are expected to be still not present. The samples were prepared by layer-by-layer thermal deposition of Cr onto a 60-thick film of Fe metal grown on a W(100) substrate followed by annealing. The crystalline order of the samples was checked by low-energy electron diffraction. In order to distinguish between contributions into normal-emission PE signal from the Cr- and the Fe-derived valence states the spectra were acquired immediately below the $3p$-$3d$ excitation threshold of Fe (the Fano antiresonance, $h\nu = 54$ eV), where the Fe $3d$ PE intensity is found to be strongly suppressed. The experiments were performed at the Russian-German beamline at BESSY II [31] using an electron-energy analyzer CLAM4 with an angular resolution of 1°. The overall system resolution was set to 130 meV FWHM. All measurements were carried out at room temperature with a base pressure in the low 10^{-9}-Pa range.

Photoemission spectra

Angle-resolved PE spectra acquired from the Cr/Fe/W structure in normal-emission geometry are shown in Fig. 8 (left). There are basically two features (A and B), which vary their binding energies with Cr coverage. Since structure A is located close to the Fermi energy, its energy shifts result in variations of the PE intensity at E_F. These BE/intensity variations have a period of ~ 2.2 ML [see Fig. 8 (right)], which is not much different as compared to the period of the short-range magnetic oscillations in Fe/Cr/Fe(100). To study origin of feature A, normal-emission PE experiments with photon-energy variation were performed. Since no dispersion of feature A was monitored, a two-dimensional character of this structure was concluded. Feature A is not a surface state or a resonance: It shows oscillating behavior with Cr thickness.

The binding energy of peak B reveals a non-monotonic behavior as well. For very low Cr coverages it drops increasing again toward the thickness of 6 ML. At 8 ML it shows minimum growing again at higher coverages.

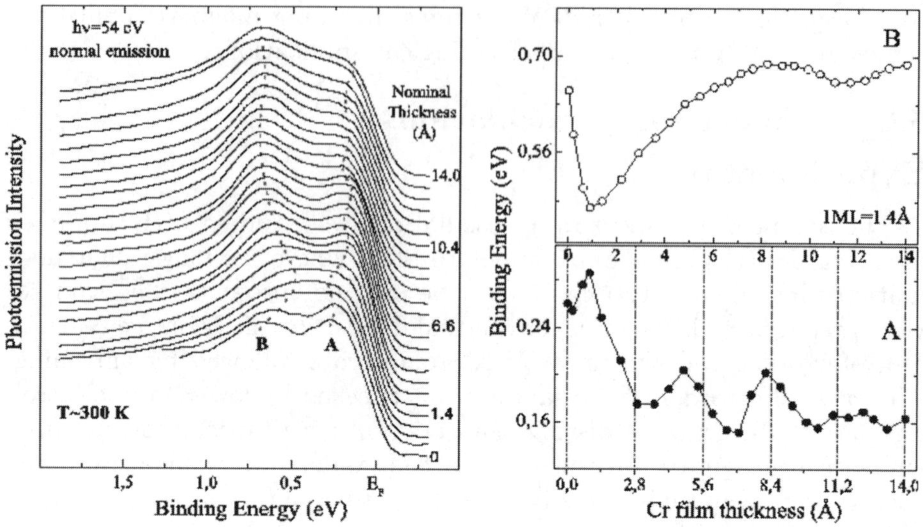

Figure 8. Normal-emission PE spectra of Cr/Fe/W(100) taken at various Cr thicknesses (left). Binding energies of features A and B as function of Cr deposition (right). Dotted vertical lines mark a 2-ML increment in increase of Cr coverage.

Electronic structure calculations

To understand the above thickness-dependent variations observed in PE experiments performed without spin resolution, the electronic structure of the system was calculated self-consistently in the framework of density functional theory using a screened-KKR Green's function method [32]. The Cr covered surface was modelled by a free Fe slab of 10 ML thickness covered on both sides with n layers of Cr (n from 1 to 10). For the potentials the atomic sphere approximation was used, nevertheless the charge density was expanded in spherical harmonics up an angular momentum of $\ell_{max} = 6$. This ensures a proper treatment of the charge relaxation at the surface. The Cr interface moment is oriented antiparallel to the adjacent Fe interface layer. To compare with the normal-emission PE spectra the spin-dependent local density of states (LDOS) of the surface Cr layer was calculated using the site diagonal part of the Green's function. Due to the 2D periodicity of the system the in-plane wave vector $\mathbf{k}_{||}$ is a good quantum number and real space properties are obtained by a Fouier transformation and integration over the 2D surface Brillouin zone. To account for the finite angular resolution of 1° the integration of the LDOS was restricted to $\mathbf{k}_{||}$ values smaller than 7% of the Brillouin zone mean radius. Furthermore the

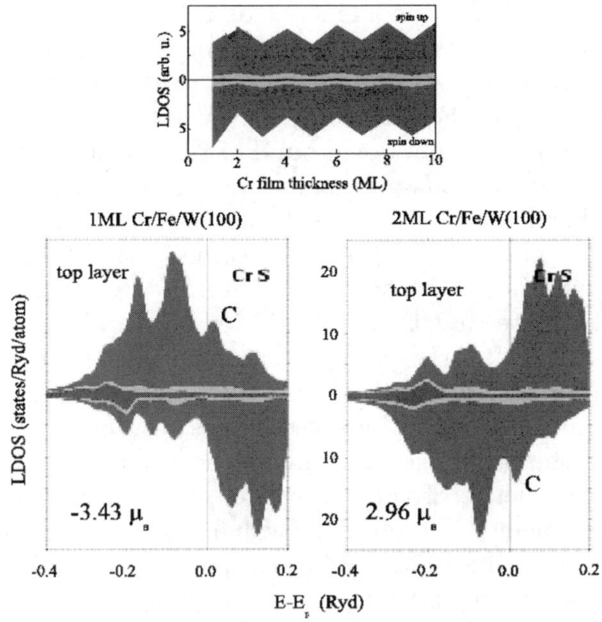

Figure 9. Spin-dependent angular-momentum resolved LDOS (bottom). Oscillations of the spun-up and the spin-down LDOS depending on thickness of Cr (top).

obtained LDOS was truncated by the Fermi-Dirac distribution function for room temperature and broadened with a Gaussian of 130-meV width according to the finite experimental energy resolution.

Discussion

Results for the spin-dependent **k**-integrated LDOS of the topmost Cr layer for thicknesses 1 and 2 ML are depicted in Fig. 9 (bottom). The *s*-, *p*-, *d*, and *f*-angular momentum character LDOS are marked by blue, green, red, and yellow, respectively. Below we will consider only *d*-like states, which reveal main contribution into the LDOS. The interface coupling of Cr with Fe is antiferromagnetic, and also the successive Cr layers carry moments of opposite sign. If Fe(100) is covered by 1 ML of Cr, this layer carries a large spin moment of -3.43 μ_B per atom. (The sign indicates AF coupling with the bulk Fe. In the following the terms spin up and spin down will be used with respect to the bulk, where spin up is defined as the majority spin.) The Cr 3*d*-spin up band is situated above E_F and thus only marginally hybridized with the Fe 3*d*-spin up band, that is almost completely occupied. As a result, the Cr 3*d*-spin-up band is quite narrow and has a steep slope just above E_F, stabilizing the large moment. In turn, the moment yields a large spin split of the Cr bands and the Fermi level is placed close to a peak of the almost occupied spin-down band (feature C). The peak gives rise to a high value of the spin-down LDOS at the Fermi energy. If Fe(100) is covered by 2 ML of Cr, the unoccupied $3d$-spin-down band of the surface layer is broader than the unoccupied Cr $3d$-spin-up in the previous case, since now a considerable hybridization with the subsurface Cr $3d$ states takes place. The result is a less steep slope above E_F leading to a reduced moment of 2.96 μ_B, a related smaller spin split and a shift of the spin-up peak above the Fermi level. (Note that the surface layer spin-up band corresponds to the spin-down band of the previous case and *vice versa*.) Hence, the total (spin up plus spin down) LDOS at E_F becomes slightly smaller.

Further increase of the Cr thickness causes continuous saturation of the variation of the magnetic moment at the value of $\sim 2.6\ \mu_B$. The latter is the consequence of the saturation of the discussed above hybridization phenomena. The obtained value is in good agreement with previously obtained values [23, 33]. Correspondingly the BE of peak C does not change anymore and constant-amplitude oscillations separately of the spin-up and the spin-down LDOS with the period of exactly 2 ML are monitored [see Fig. 9 (top)]. It is expected that these oscillations, which are due to the AF interaction between neighboring Cr layers, are responsible for the 2-ML short-range oscillations of the magnetic coupling between marginal Fe layers in Fe/Cr/Fe structures. Note that the total LDOS does not vary anymore at Cr coverages higher than about 3 ML.

Therefore these spin-resolved **k**-integrated LDOS modulations cannot explain the behavior of our non-spin-resolved experimental spectra particularly close to the Fermi energy (feature A). Moreover the angle-resolved PE data taken in the normal-emission geometry should be compared to results of LDOS calculations restricted to the **k** region close to $\mathbf{k}_{\parallel} = 0$. The corresponding **k**-resolved LDOS results for topmost layers of Cr in the cases of 2-, 4-, 6-, and 8-ML Cr systems are shown in Fig. 10. To analyze the character of the electron eigenstates the probability amplitudes of the states in the center of the Brillouin zone were calculated. For the states close to the Fermi level we find a strong localization effect that means all states are localized in the direction perpendicular to the surface inside the Fe or the Cr layer. For the Cr states with a probability amplitude mainly concentrated in the Cr layer two types of states can be distinguished: surface and quantum well states. In Fig. 10 the surface state is always located at E_F and seen in the spin-up LDOS (feature D). In difference to the surface state the quantum well states, which are monitored in this figure in the spin-down LDOS, change

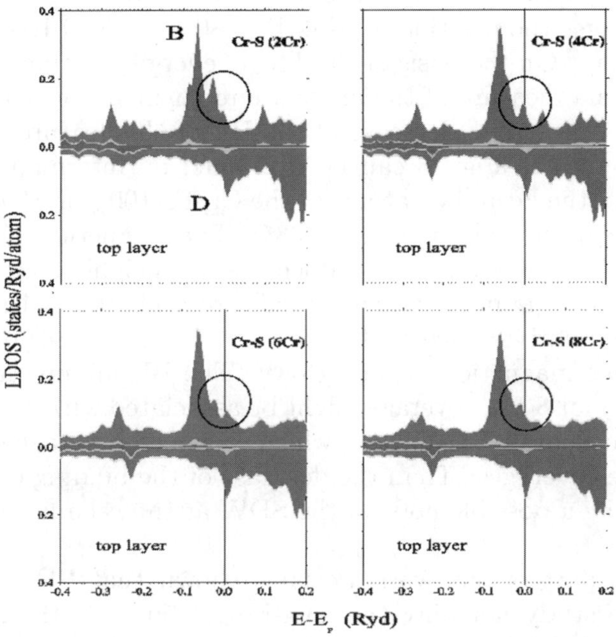

Figure 10. **k**-resolved LDOS for 2 ML, 4 ML, 6 ML, and 8 ML of Cr on Fe/W(100).

their positions periodically crossing the Fermi energy with Cr coverage (marked by circles in the figure). Thereby the period of these crossings, which are displayed in the spin-integrated **k**-resolved LDOS as variations of intensities at E_F, is very close to the one observed in our PE experiment. By this reason we assign the PE intensity oscillations of peak A in our data to the QWS behavior. Our study is a first observation of the QWS in <100> directions in thin films of chromium. It is anticipated that the QWS mechanism can contribute into the short-range regime of the magnetic coupling in Fe/Cr/Fe systems. Its contribution, however, is weaker than that of the bare AF interaction, since the QWS LDOS is approximately one order of magnitude smaller than the LDOS associated with the AF coupling. One may speculate that the aliasing of the QWS and the AF modulations of the LDOS can cause oscillations with a periodicity close to that of the long-range oscillations of the magnetic coupling in Fe/Cr/Fe multiplayer structures.

Feature B in Fig. 8 can be assigned to the main LDOS peak (Fig. 10), which is located at about 1.2 eV (0.088 Ryd) BE for the system 1 ML of Cr on Fe(100) [not shown]. As obtained from our calculations the binding energy of this peak follows changes of the magnetic moment of Cr. With decrease of the latter, peak B shifts toward the Fermi energy.

For the 2-ML system it is found at 0.85 eV (0.063 Ryd). Beginning from 3-4 ML coverages the position of peak B [0.80 eV (0.059 Ryd)] does not change anymore. On the basis of the above, energy location of feature B can be used as a measure of the magnetic moment in the surface region of Cr layers. The strong decrease of the BE of this feature at very low Cr coverages seen in Fig. 8 can be explained by intermixing at room temperature of the Fe and Cr atoms at the Cr/Fe(100) interface observed also by other experimental techniques [34]. The magnetic moment at the intermixed interface is strongly reduced, since neighboring Fe and Cr atoms trend to align antiferromagnetically to each other. Upon further Cr deposition, relative concentration of Fe in the surface region decreases and the surface magnetic moment grows. The BE minimum of peak B observed at about 8 ML coverage might be associated with a constitution of the incommensurate SDW state, which is expected to take place in the range of these coverages. Than the decrease of the binding energy could be explained by a possible node of the SDW at the vacuum boundary of the Cr layer.

In the last part of our work we have shown that PE is an appropriate tool to study not only the long-range, but also the short-range oscillations in thin films of Cr. The short-range photoemission intensity modulations at the Fermi energy are related to the quantum-well states, which were for the first time observed in <100> directions in Cr(100) layers. Possible contributions of the QWS into the short-range and the long-range magnetic coupling between marginal layers in Fe/Cr/Fe system were discussed. It was found that the binding energy of the main peak in the **k**-resolved normal-emission LDOS of the topmost layer of Cr can be used to follow the magnetic moments at the surfaces of Cr systems.

5.3 Acknowledgments

This work was supported by the DFG, SFB 463 TP B16, the BMBF (project 05-SF80D1/4) and by the bilateral project "Russian-German Laboratory at BESSY II".

References

[1] E. Fawcett, Rev. Mod. Phys. **60**, 209 (1988).

[2] A.W. Overhauser, Phys. Rev. **128**, 1437 (1962); S.A. Werner, A. Arrott, and H. Kendrick, Phys. Rev. **155**, 528 (1967).

[3] R.H. Victora and L.M. Falicov, Phys. Rev. B **31**, 7335 (1965); Y. Sakisaka, T. Komeda, M. Onchi, H. Kato, S. Suzuki, K. Edamoto, and Y. Aiura, Phys. Rev. B **38**, 1131(1988).

[4] M. van Schilfgaarde and W.A. Harrison, Phys. Rev. Lett. **71**, 3870 (1993).

[5] P. Bruno and C. Chappert, Phys. Rev. Lett. **67**, 1602 (1991); D.D. Koelling, Phys. Rev. B **50**, 273 (1994).

[6] M.D. Stiles, Phys. Rev. B **48**, 7238 (1993); M.D. Stiles, Phys. Rev. B **54**, 14679 (1996).

[7] Dongqi Li, J. Pearson, S.D. Bader, E. Vescovo, D.-J. Huang, P.D. Johnson, and B. Heinrich, Phys. Rev. Lett. **78**, 1154 (1997).

[8] K.-F. Braun, S. Flsch, G. Meyer, and K.-H. Rieder, Phys. Rev. Lett. **85**, 3500 (2000).

[9] J.S. Dodge, A.B. Schumacher, J.-Y. Bigot, D.S. Chemla, N. Ingle, and M. R. Beasley, Phys. Rev. Lett. **83**, 4650 (1999); I.I. Mazin, D.J. Singh, and Claudia Ambrosch-Draxl, Phys. Rev. B **59**, 411 (1999).

[10] M.N. Baibich, J.M. Broto, A. Fert, F. Nguyen Van Dau, F. Petroff, P. Eitenne, G. Creuzet, A. Friederich, and J. Chazelas, Phys. Rev. Lett. **61**, 2472 (1988); G.A. Prinz, J. Magn. Magn. Mater. **200**, 57 (1999); O.J. Monsma, J.C. Lodder, Th.J.A. Popma, and B. Dieny, Phys. Rev. Lett. **74**, 5260 (1995).

[11] S.S.P. Parkin, N. More, and K. P. Roche, Phys. Rev. Lett. **64**, 2304 (1990).

[12] T.G. Walker, A.W. Pang, H. Hopster, and S.F. Alvarado, Phys. Rev. Lett. **69**, 1121 (1992).

[13] J. Unguris, R.J. Celotta, and D.T. Pierce, Phys. Rev. Lett. **69**, 1125 (1992); D.T. Pierce, Joseph A. Stroscio, J. Unguris, and R.J. Celotta, Phys. Rev. B **49**, 14564 (1994).

[14] S.T. Purcell, W. Folkerts, M.T. Johnson, N.W.E. McGee, K. Jager, J. aan de Stegge, W.B. Zeper, W. Hoving, and P. Grnberg, Phys. Rev. Lett. **67**, 903 (1991).

[15] M.A. Ruderman and C. Kittel, Phys. Rev. **96**, 99 (1954).

[16] A.M.N. Niklasson, J.M. Wills, and L. Nordstrm, Phys. Rev. B **63**, 104417 (2001).

[17] A.M.N. Niklasson, B. Johansson, and L. Nordstrm, Phys. Rev. Lett. **82**, 4544 (1999).

[18] K. Hirai, Phys. Rev. B **59**, R6612 (1999).

[19] J. Schfer, E. Rotenberg, G. Meigs, S. D. Kevan, P. Blaha, and S. Hfner, Phys. Rev. Lett. **83**, 2069 (1999).

[20] N. Nakajima, O. Morimoto, H. Kato, and Y. Sakisaka, Phys. Rev. B **67**, R041402 (2003).

[21] further details will be published by F. Schiller, D.V. Vyalikh, V.D.P. Servedio, and S.L. Molodtsov.

[22] D.V. Vyalikh, Manuel Richter, Yu.S. Dedkov, and S.L. Molodtsov, J. Magn. Magn. Mater., in print; further details will be published by D.V. Vyalikh, P. Zahn, Manuel Richter, Yu.S. Dedkov, and S.L. Molodtsov.

[23] D. R. Grempel, Phys. Rev. B **24**, 3928 (1981).

[24] S.V. Halilov, E Tamura, D Meinert, H Gollisch, and R Feder, J. Phys.: Condens. Matter **5**, 3859 (1993).

[25] S. Asano and J. Yamashita, J. Phys. Soc. Jpn. **23**, 714 (1967).

[26] S.L. Molodtsov, F. Schiller, S. Danzenbcher, Manuel Richter, J. Avila, C. Laubschat, and M. C. Asensio, Phys. Rev. B **67**, 115105 (2003).

[27] F. Reinert, D. Ehm, S. Schmidt, G. Nicolay, S. Hfner, J. Kroha, O. Trovarelli, and C. Geibel, Phys. Rev. Lett. **87**, 106401 (2001).

[28] A. Lanzara, P.V. Bogdanov, X.J. Zhou, S.A. Kellar, D.L. Feng, E.D. Lu, S. Uchida, H. Eisaki, A. Fujimori, K. Kishio, J.-I. Shimoyama, T. Noda, S. Uchida, Z. Hussain, and Z.-X. Shen, Nature **412**, 510 (2001); A.D. Gromko, A.V. Fedorov, Y.-D. Chuang, J.D. Koralek, Y. Aiura, Y. Yamaguchi, K. Oka, Yoichi Ando, and D.S. Dessau, cond-mat/0202329; A.D. Gromko, Y.-D. Chuang, A.V. Fedorov, Y. Aiura, Y. Yamaguchi, K. Oka, Yoichi Ando, and D.S. Dessau, cond-mat/0205385; A. Koitzsch, S.V. Borisenko, A.A. Kordyuk, T.K. Kim, M. Knupfer, J. Fink, H. Berger, and R. Follath, cond-mat/0305380.

[29] S.H. Lui, Phys. Rev. B **2**, 2664 (1970).

[30] M. Li and N.X. Chen, Z. Phys. B **100**, 169 (1996); G. Simonelli, R. Pasianot, and E. J. Savino, Phys. Rev. B **55**, 5570 (1997).

[31] S.I. Fedoseenko, D.V. Vyalikh, I.E. Iossifov, R. Follath, S.A. Gorovikov, R. Pttner, J.-S. Schmidt, S.L. Molodtsov, V.K. Adamchuk, W. Gudat, and G. Kaindl, Nucl. Instr. and Meth. in Phys. Res. A **505**, 718 (2003).

[32] L. Szunyogh, B. jfalussy, and P. Weinberger, Phys. Rev. B **51**, 9552 (1995); R. Zeller, P. H. Dederichs, B. jfalussy, L. Szunyogh, and P. Weinberger, Phys. Rev. B **52**, 8807 (1995); N. Papanikolaou, J. Opitz, P. Zahn, and I. Mertig, Phys. Rev. B **66**, 165441 (2002).

[33] C.L. Fu and A.J. Freeman, Phys. Rev. B **33**, 1755 (1986).

[34] A. Davies, Joseph A. Stroscio, D.T. Pierce, and R.J. Celotta, Phys. Rev. Lett. **76**, 4175 (1996); D. Venus and B. Heinrich, Phys. Rev. B **53**, R1733 (1996).

THE ROLE OF HYDRATION AND MAGNETIC FLUCTUATIONS IN THE SUPERCONDUCTING COBALTATE

M.D. Johannes, D.J. Singh
Center for Computational Material Science
Naval Research Laboratory
Washington, D.C. 20375

Abstract We report electronic structure calculations within density functional theory for the hydrated superconductor $Na_{1/3}CoO_2 1.33H_2O$ and compare the results with the parent compound $Na_{0.3}CoO_2$. We find that intercalation of water into the parent compound has little effect on the Fermi surface outside of the predictable effects expansion, in particular increased two-dimensionality. This implies an intimate connection between the electronic properties of the hydrated and unhydrated phases. Additional density functional calculations are used to investigate the doping dependence of the electronic structure and magnetic properties in hexagonal Na_xCoO_2. The electronic structure is highly two dimensional, even without accounting for the structural changes associated with hydration. At the local spin density approximation level, a weak itinerant ferromagnetic state is predicted for all doping levels in the range $x = 0.3$ to $x = 0.7$, with competing but weaker itinerant antiferromagnetic solutions. Comparison with experiment implies substantial magnetic quantum fluctuations. Based on the simple Fermi surface and the ferromagnetic tendency of this material, it is speculated that a triplet superconducting state analogous to that in Sr_2RuO_4 may exist here.

Keywords: Hydrated superconductor, magnetic quantum fluctuations, triplet superconductivity

Introduction

During the past year, the discovery of likely unconventional superconductivity in $Na_xCoO_2 \cdot yH_2O$ ($x \sim 1/3$, $y \sim 4/3$) and the unusual magnetotransport properties of Na_xCoO_2 (x$\sim$2/3), have focused attention on these materials and the connection between them.[1] In fact, lay-

S. Halilov (ed.), Physics of Spin in Solids: Materials, Methods and Applications, 85–99.

ered cobalt oxide materials have lately been the subject of considerable fundamental and practical interest for several reasons. Li_xCoO_2 is an important cathode material for lithium batteries. In that context, the interplay between the transition metal-oxygen chemistry, the Co mixed valence and magnetism are important ingredients in the performance of the material. [2] Layered cobaltates, A_xCoO_2 also form for A=Na and K, but in more limited concentration ranges. [3] Single crystals of Na_xCoO_2, with nominal x=0.5 were investigated by Terasaki and co-workers. [4] Remarkably, they found that even though the material is a good metal, it also has a large thermopower of approximately 100 μV/K at room temperature. This was the first time that an oxide showed promise of matching the thermoelectric performance of conventional heavily doped semiconductors for thermoelectric power conversion. Interest in modifications of this material to minimize its thermal conductivity for thermoelectric applications led to the discovery that similar anomolously high thermopowers were present in other materials with hole doped CoO_2 layers [5–7], especially so-called misfit compounds in which the intercalating Na is replaced by more stable rocksalt like oxide blocks. [8–12] This demonstrates that the exceptional thermopower of metallic Na_xCoO_2 is not essentially related to the details of the intercalating layer.

Theoretical studies, stimulated by these discoveries, emphasized both the band-like nature of the material, [13, 14] consistent with its good metallic properties, the proximity to magnetism, and possible renormalizations related to a magnetic quantum critical point, [15] the proximity to charge ordering (which is seen at some doping levels), [16] and possible strongly correlated electron physics [17, 18]. Intriguingly, both band structure and strong correlated models (*i.e.* the Heikes model) are able to explain the high thermopowers. [19, 20] Recently, it was demonstrated by magnetotransport measurements that the thermopower at $x \sim 0.68$ is strongly reduced in magnetic field with a universal scaling law [21] showing that spin entropy underlies the high thermopower and thus again emphasizing the role of magnetic fluctuations in the system as well as possible strong correlated electron physics. [21, 22] Indeed, some of the misfit compounds are in fact magnetic, with ferromagnetic ground states. [23]

The last few years have seen the discovery of a number of novel unconventional superconductors associated with magnetic phases. Focusing on triplet (or likely triplet) superconductors, these include UGe_2 ($T_c \sim 1$K), [24] URhGe ($T_c \sim 0.25$K), [25] and $ZrZn_2$ ($T_c \sim 0.3$K), [26, 27] where ferromagnetism coexists with superconductivity, and Sr_2RuO_4 ($T_c \sim$ 1.5K), [28, 29] which has a paramagnetic Fermi liquid normal state, but is "near" magnetic phases. Although the exact pairing mechanism has not been established in these materials, it is presumed that spin fluctuations are involved, most probably the quantum critical fluctuations in the materials with co-existing ferromagnetism and superconductivity. [30–36] In Sr_2RuO_4, strong nesting related antiferromagnetic spin fluctuations are found in local density approximation (LDA) calculations and experiment. [37, 38] In addition ferromagnetic fluctuations, which have recently been observed in the parent compound at a doping level of 0.75 [39], may also be present, and if so, these would favor a triplet superconducting state. [40]

Takada and co-workers recently showed that Na_xCoO_2 can be readily hydrated to form $Na_xCoO_2 \cdot yH_2O$. This material has the same CoO_2 layers, but with a considerably expanded c axis, which accomodates the intercalating water and Na. In this material, $x \sim 0.3$, is lower than the range readily formed in Na_xCoO_2. Remarkably, Takada and co-workers found that $Na_xCoO_2 \cdot yH_2O$ is a superconductor with $T_c \sim 5$K. [41, 42] The nearness to magnetism and possible strong correlations immediately lead to suggestions of unconventional superconductivity in this material, beginning with the discovery paper of Takada, as well as discussions of the role of water in producing the superconductivity. Scenarios that have been advanced include no role at all, screening Na disorder, preventing competing charge ordered states, modifying the doping level via unusual chemistry, enhanced two-dimensionality, and others. [43–49]

Since superconductivity is fundamentally an instability of the metallic Fermi surface, a first step is to understand the relationship of the electronic structure of $Na_xCoO_2{\cdot}yH_2O$ with its unhydrated parent Na_xCoO_2. We present density functional based bandstructure calculations using the linearized augmented planewave (LAPW) method [50, 51] of $Na_{1/3}CoO_2 \cdot 4/3H_2O$ with both Na ions and water molecules explicitely included (no virtual crystal approximation is made). In the first part of this manuscript, we show that, from an electronic structure point of view, the hydrated and unhydrated compounds are identical, aside from structural effects due to the expansion of the c-axis. With the knowledge that the electronic structure of the parent compound is likely to reflect that of the superconducting compound, we proceed in the second part of the manuscript to investigate Na_xCoO_2 in terms of its doping dependencies, magnetic properties, and possible quantum critical fluctuations. Based on the two dimensional $3d$ transition metal oxide structural motif, there are speculations that the superconductivity may be related to that of the cuprate high-T_c superconductors. Here an alternate possibility is discussed, that is the connection with the triplet superconductors mentioned above.

6.1 Hydrated and Unhydrated Band Structures

The exact structure of the hydrated compound has yet to be conclusively resolved. So far, all experimental data [52–54, 42, 43] indicates that it belongs to the hexagonal symmetry $P6_3/mmc$ (# 194), but refinements of the water molecule positions, Na ion positions, and apical oxygen heights vary. Lynn *et al* [52] find that the Na ions are displaced compared to the unhydrated parent compound and are surrounded by H_2O molecules with basically the same structure as D_2O ice. However, other neutron diffraction studies show that [55], even below the freezing point of water, there may be no static position for the water molecule as a whole, emphasizing disorder.

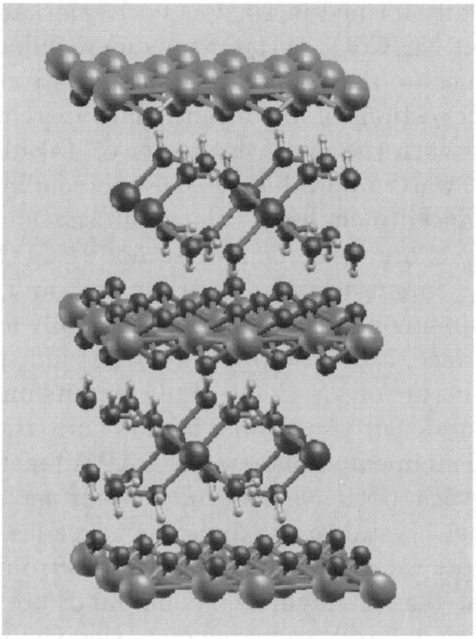

Figure 1. The tripled, hydrated structure corresponding to the superconductor. The green and red planes are Co-O, dark blue ions are Na, and the red and light blue molecules are water. Bonds between the Na molecules are drawn to emphasize the four-fold coordination. One of the H ions is located 1.73 Åfrom an O ion in the plane above it, the other position is determined by preserving the H_2O bond angle and maximizing the distance between any two H ions

The structure used for our calculation was based in large part on the neutron and x-ray diffraction results of Ref. [53] and is shown in Fig. 2. To achieve this structure fully occupied shifted Na ($6h$) and H_2O ($24l$) sites were assumed and then ions/molecules were systematically removed according to coordination and bonding rules until the observed proportions were obtained. The bond angles are held to 109° and the O-H bond distance to 0.99 Å. We employed this structural configuration by tripling the formula unit of the parent compound $Na_{2x}Co_2O_4$ (already doubled to account for both Co-O planes), expanding the c-lattice to its reported value [53] of 37.1235 a.u. and adding four H_2O molecules for each Na ion, resulting in a formula unit with integer values of all constituent atoms: $Na_2Co_6O_{12}8H_2O$. This eliminated the need for the virtual crystal approximation, allowing us to take the possible effects of Na ordering into account, which we find *a posteriori* to be unimportant for the CoO_2 derived electronic structure, based on comparison with previous virtual crystal results. The water molecules were oriented with H ions pointing away from the Na ion and toward the Co-O plane. Since the Na and H_2O sites are only partially occupied in the $P6_3/mmc$ symmetry, we required a different space group for computation. Our structure has the considerably lowered $P2_1/m$ symmetry (#11), but remains pseudohexagonal with lattice vectors of length $\sqrt{3}$a such that the planar area of the unit cell is tripled. In other words, we include the local structure and coordination, but not long range disorder in the Na H_2O layers, yielding a lower average symmetry. However, electronic structure around the Fermi level is hexagonal to a high precision, indicating that scattering from disorder in the Na H_2O layers is weak. This may be important for superconductivity considering the pair breaking effect of scattering in unconventional superconductors. The apical oxygen height was relaxed to its optimal position 1.81 a.u. above the Co plane. We oriented each water molecule with both H ions pointing away from the Na ion and with one as close as possible to an O in the Co-O plane. The present local density approximation calculations were done using the LAPW method as implemented in the WIEN2k code with well converged basis sets employing an Rkmax of 4.16, sphere radii of 1.86 (Co), 1.6 (O), 2.0 (Na), and 1.0 and 0.88 for the O and H respectively of water. The water molecules were treated using LAPW basis functions, whereas all other ions were treated with an APW + LO basis set. Additional local orbitals were added for Co and Na p-states and O s-states.

The most important bands in the conventional hexagonal Brillouin zone (BZ) of the unhydrated parent compound are four E_g' and two A_{1g} Co-derived bands near the Fermi energy. Our expanded hexagonal unit cell results in a BZ one third the volume of the original and rotated by an

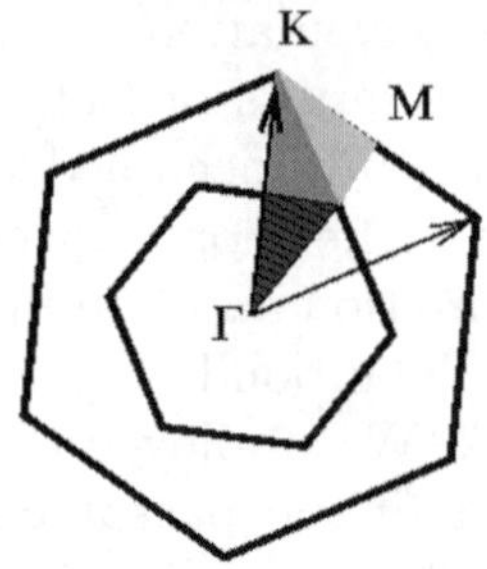

Figure 2. The bands along the Γ-M in the small zone are formed by the folding of the blue triangle down onto the green and then again onto the red in the irreducible Brillouin zone. The symmetry points marked are those of the larger zone, those of the smaller zone are easily identifiable by analogy

angle of 30°. The rotation and expansion of the original BZ necessitates a double downfolding process as illustrated in Fig. 2, and needs to be remembered when comparing our band structure and prior results. [13, 16].

To clarify the similarities and differences between the hydrated and unhydrated structures, we performed a second calculation in a similar unit cell, neglecting the water molecules. We found that artificially expanding the c-axis with a vacuum produced unphysical and highly dispersive bands. The c-axis in our comparison calculation was fixed at its lower unhydrated value of 20.4280 a.u. for this reason.

Fig. 3 shows both bandstructures on the same energy scale, each centered around its respective Fermi energy. The difference in c-axis parameter is reflected in the Γ - A distance which is nearly twice as big in the parent compound. An inspection of the bands crossing and just above the Fermi energy reveals a somewhat greater splitting in the unhydrated compound than in the superconducting compound, but a nearly identical overall band dispersion. Bands containing water character are determined to be at least 0.2 Ryd below the Fermi energy by looking at the projected atomic character of each eigenvalue. The observable increase in splitting can be attributed to interplanar coupling which is substantially suppressed when the c-axis expands to accomodate water. Thus, the sole effect of the water on the electronic structure is to collapse the two (nearly) concentric Fermi surfaces of the unhydrated compound until they are practically a single degenerate surface in the hydrated compound. While this collapse may be important, it is a purely structural effect achieved by the forced separation of Co-O planes and is unrelated to the specific chemical composition of water. This shows that the water itself, at least in this or similar structural configurations, is

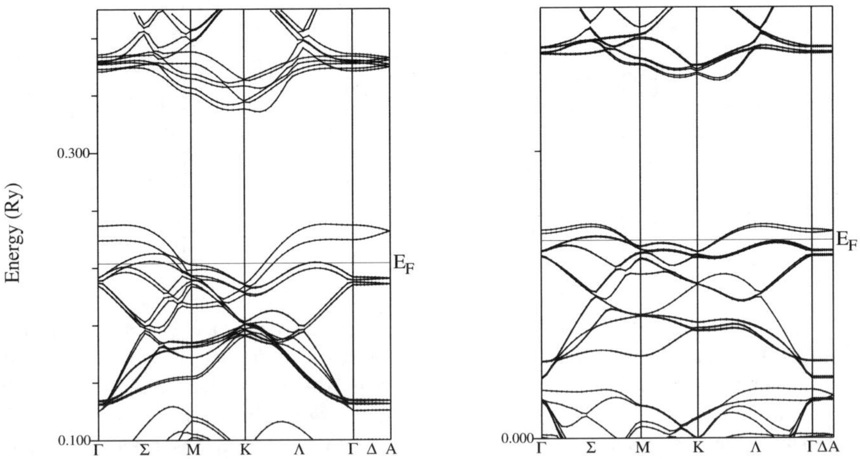

Figure 3. A comparison of the hydrated compound (lower panel) with its expanded c lattice parameter and an unhydrated compound with identical dimensions with the exception of the c-axis which remains at the unhydrated value. The differences in dispersion between the two structure is completely attributable inter-planar interaction which is reduced by hydration.

completely irrelevant to the electronic structure of $Na_{0.33}CoO_{2}1.33H_{2}O$ and that the Fermi surface is insensitive to its presence. This result does not depend on the specific position and orientation of the water. We calculated another bandstructure, with water positions based loosely on the ice-like model [52] and obtained an identical result for the partially filled Co bands.

6.2 Quantum Critical Fluctuations

Our calculations in the previous section lead us to believe that results obtained for the parent, or unhydrated, cobaltate will retain their validity in the superconducting compound and that electronic structure properties of the system can be accurately obtained without employing the full, extensive structure. Here, well converged LDA calculations are reported for $Na_{x}Co_{2}O_{4}$ for x=0.3,0.5,0.7. In addition, calculations are reported for a strained lattice corresponding to the structure reported for superconducting $Na_{x}CoO_{2} \cdot yH_{2}O$, but neglecting the intercalating water. The calculations were done using the general potential linearized augmented planewave method with local orbitals, [51, 56] as described

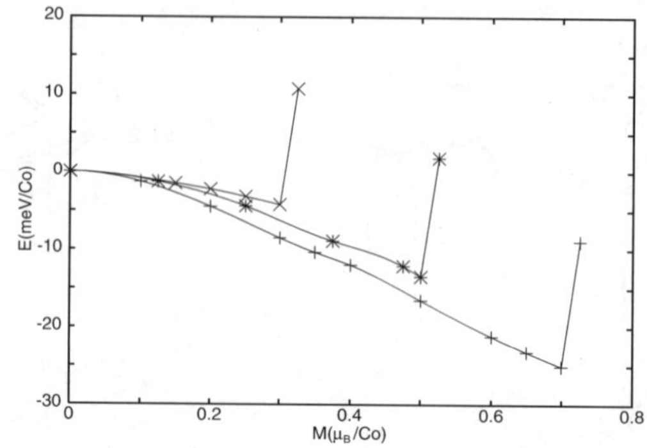

Figure 4. LDA fixed spin moment energy as a function of constrained spin magnetization of Na_xCoO_2 on a per Co atom basis for x=0.7 (×), x=0.5 (*) and x=0.3 (+). The curves are spline interpolations as a guide to the eye. Note the breaks at 0.3, 0.5 and 0.7 μ_B for x=0.7, 0.5 and 0.3, respectively. These correspond to the band gap between crystal field split Co d manifolds. The calculations were done keeping the structure fixed and varying the Na site occupation via the virtual crystal method (see text).

in Ref. [57], except that better Brillouin zone samples, corresponding to a minimum special **k**-points mesh of 16 × 16 × 2 in the hexagonal zone are used here to obtain convergence of the magnetic energies. As discussed in Ref. [57], a virtual crystal method is used to account for the partially occupied Na site.

Although there is some hybridization, the valence band structure of $Na_{0.5}CoO_2$ consists of three manifolds of bands separated by gaps – a lower lying occupied O $2p$ derived manifold, followed by Co t_{2g} and e_g manifolds. [57] As expected from ionic considerations the Fermi energy, E_F lies near the top of the t_{2g} manifold, which contains 0.5 holes per Co ion. Because of the actual axial site symmetry, the t_{2g} manifold can be regarded as consisting of two-fold (also labeled e_g) and one-fold (labeled a_g) crystal field states. These overlap, but the top of the t_{2g} manifold is primarily of a_g character in $Na_{0.5}CoO_2$, with the result that the band structure near E_F can be roughly viewed as consisting of one band per Co ion with a filling of 3/4 (1/2 hole per Co). The present LDA calculations for the electronic structure of $Na_{0.7}CoO_2$ and $Na_{0.3}CoO_2$ follow this picture. The Fermi surfaces consist of simple rounded hexagonal cylinders, centered on the $\Gamma - A$ line, and additional small sections. These calculations were done holding the crystal structure fixed at that of $Na_{0.5}CoO_2$ and varying the Na occupancy in the virtual crystal. The dominant a_g character at the top of the t_{2g} manifold is, however, lost for the strained lattice, discussed below.

The magnetic properties are similar for the various Na contents in the range $x = 0.3$ to $x = 0.7$. Results of fixed spin moment constrained LDA calculations are shown in Fig. 4. In particular, itinerant ferromagnetism is found. In each case, the energy decreases with magnetization until a magnetization at which the band edge is reached in the majority spin. Then the energy increases rapidly reflecting the crystal field induced gap between the t_{2g} and e_g manifolds.

Thus, independent of x in this range, the LDA predicts a ferromagnetic ground state, with a spin moment per Co equal to the number of holes ($p = 1 - x$) and a half metallic band structure (here we refer to the hole concentration as the concentration of holes in the t_{2g} manifold; without any Na, p=1; Na electron dopes the sheets, which leads to a reduction in p). The fixed spin moment curves show a shape crossing over from parabolic at low moment to more linear as the band edge is approached. As may be seen, the shapes and initial curvatures for the different doping levels are roughly similar. The trend towards slightly weaker initial curvatures at higher Na concentration is possibly an artifact due to the fixed crystal structure used in the present calculations. It reflects increasing hybridization (increasing band width and decreasing density of states) as charge is added to the CoO_2 planes. In reality, the lattice would be expected to expand, perhaps compensating this trend. In any case, for this range of x, ferromagnetism with a magnetic energy of approximately, $E(FM) \approx -50p$ in meV/Co and spin moment $M(FM) = p$ in μ_B/ Co is found. As mentioned, calculations were also done for a strained cell with the structure of the superconducting sample, but without H_2O. These calculations were done for x=0.5 and x=0.35, the latter corresponding to the experimentally determined doping level. In both cases, the LDA predicted a ferromagnetic state. The magnetic energies were $E(FM)$ =-20 and -27 meV/Co for x=0.5 and x=0.35, re-

Table 1. LSDA spin magnetizations and energies for Na_xCoO_2. All quantities are on a per Co basis. Energies are in meV, spin moments are in μ_B, FM denotes ferromagnetic and AF denotes the partially frustrated nearest neighbor AF configuration discussed in the text. M is the total spin magnetization, and m is the magnetization inside the Co LAPW sphere, radius 1.95 Bohr. Negative energies denote instabilities of the non-spin-polarized state.

	E(FM)	M(FM)	m(FM)	E(AF)	m(AF)
x=0.3	-25.	0.70	0.56	-9.	0.36
x=0.5	-13.	0.50	0.41	-3.	0.21
x=0.7	-4.	0.30	0.25	$\geq$-1.	0.04

spectively. Thus the behavior is similar, but the magnetic energies are somewhat larger in magnitude.

LDA calculations were also done for an antiferromagnetic configuration with the unit cell doubled along one of the in-plane lattice vectors. Thus, within a Co plane, each Co ion has four opposite spin nearest neighbors and two like spin nearest neighbors. At all three doping levels investigated an antiferromagnetic instability was found, but this instability is weaker than the ferromagnetic one. Details of the LDA moments and energies are given in Table 1. Essentially, the energy of the antiferromagnetic configuration examined tracks the ferromagnetic energy at a value $\sim 1/4$ as large in this range of x. [58]

The LDA generally provides a good description of itinerant ferromagnetic materials. It is known to fail for strongly correlated oxides where on-site Coulomb (Hubbard) repulsions play an important role in the physics. In such cases, the LDA underestimates the tendency of the material towards local moment formation and magnetism. Here, the LDA is found to predict ferromagnetic ground states for materials that are paramagnetic metals in experiment. While materials for which the LDA substantially overestimates the tendency towards magnetism are rare, a number of such cases have been recently found. These are generically materials that are close to quantum critical points, and include Sc_3In, [59] $ZrZn_2$, [60] and $Sr_3Ru_2O_7$ (Ref. [61]). $Sr_3Ru_2O_7$ displays a novel metamagnetic quantum critical point, [62] while, as mentioned, $ZrZn_2$ shows coexistence of ferromagnetism and superconductivity.

Density functional theory is in principle an exact ground state theory. It should, therefore, correctly describe the spin density of magnetic systems. However, common approximations to the exact density functional theory, such as the LDA, neglect Hubbard correlations beyond the mean field level, yielding the underestimated magnetic tendency of strongly Hubbard correlated systems. Overestimates of magnetic tendencies, especially in the LDA are very much less common. Another type of correlations that is missed in these approximations are quantum spin fluctuations. This is because the LDA is parameterized based on electron gases with densities typical for atoms and solids. However, the uniform electron gas is very far from magnetism in this density range. In solids near quantum critical points, the result is an overestimate of the magnetic moments and tendency toward magnetism (*i.e.* misplacement of the position of the critical point) due to neglect of the quantum critical fluctuations. [63, 64]

The present results for Na_xCoO_2 show a weak ferromagnetic instability that is robust with respect to doping and structure (note the instability for the strained lattice). Based on this, and the experimentally observed renormalized paramagnetic state, it seems likely that Na_xCoO_2 is subject to strong ferromagnetic quantum fluctuations of this type, and that these are the reason for the disagreement between the LDA and experimental ground states.

The effects of such quantum fluctuations can be described on a phenomenological level using a Ginzburg-Landau theory, in which the magnetic properties defined by the LDA fixed spin moment curve are renormalized by averaging with an assumed (usually Gaussian) function describing the beyond LDA critical fluctuations. [65, 66] Although a quantitative theory allowing extraction of this function from first principles calculations has yet to be established, one can make an estimate based on the LDA fixed spin moment curves as compared with experiment. In particular, Na_xCoO_2 shows a disagreement between the LDA moment and experiment equal to $p = 1 - x$, and has a very steeply rising LDA energy for moments larger than p. Thus one may estimate an r.m.s. amplitude of the quantum fluctuations of $\xi \approx \alpha p$ in μ_B, with $1/2 < \alpha < 1$, and most likely closer to 1. These are large values *c.f.* $ZrZn_2$. It is therefore tempting to associate the superconductivity of $Na_xCoO_2 \cdot yH_2O$ with ferromagnetic quantum critical fluctuations. Considering the simple 2D Fermi surface, which consists of rounded hexagonal cylinders plus small sections, [57] and the ferromagnetic fluctuations, a triplet state like that originally discussed for Sr_2RuO_4 (Ref. [29]) seems plausible. Speculations about the ingredients in a spin fluctuation mediated triplet superconducting scenario are now given based on the calculated results.

Within a spin fluctuation induced pairing approach analogous to that employed for Sr_2RuO_4 the key ingredient is the integral over the Fermi surface of the **k**-dependent susceptibility with a function of the assumed triplet symmetry, [30, 31, 40] *i.e.* in the simplest case, $\mathbf{k} \cdot \mathbf{k}'/kk'$. For a Fermi surface in the shape of a circular cylinder, radius k_F, the needed integral is proportional to $\int_0^{2\pi} d\theta cos(\theta) V(2k_F sin(\theta/2))$, where $V(k)$ is the assumed pairing interaction. In any case, for a smooth variation of the spin fluctuations with k and a maximum at k=0 (ferromagnetic), the integral is roughly proportional to k_F times the variation of V from $k = 0$ to $k = 2k_F$. This latter variation depends on the detailed shape of $V(k)$, but may be expected to cross over from being proportional to k_F^2 for small k_F to proportional to k_F for larger k_F. Neglecting small Fermi surface sections, k_F varies as $p^{1/2}$. One possibility for $V(k)$ is a function smoothly going from a finite value at $k = 0$ to near zero at the zone boundary (reflecting the rather weak antiferromagnetic instability relative to the ferromagnetic), with a size at $k = 0$ given by the LDA ferromagnetic energy ($\propto p^2$) or alternately a Hund's exchange coupling (p independent) times ξ ($\propto p$).

Within such a p-wave scenario it would be quite interesting to measure the variation of the superconducting properties of $Na_xCoO_2 \cdot yH_2O$ as a function of doping level. The above arguments imply a substantial model dependent variation up to the level where proximity to the critical point suppresses T_c, with the implication that still higher values of T_c may be obtained. It should be stated that unconventional superconductivity is in general more sensitive to scattering than s-wave superconductivity and so the effect of scattering due to Na and H_2O disorder may be significant, and besides it should be emphasized that the mechanism and superconducting symmetry of $Na_xCoO_2 \cdot yH_2O$ have yet to be established, and in fact, even conventional electron-phonon superconductivity competing with spin fluctuations has not been excluded.

Summary and Open Questions

We have shown, by explicitely including water in an LAPW calculation, that the effect of the water in $Na_{1/3}CoO_2 1.33H_2O$ is overwhelmingly structural and imperceptibly electronic. The bandstructures of the hydrated and unhydrated compounds differ only through suppression of inter-planar coupling. The resulting decrease in bandsplitting may have relevance to superconductivity, but the same effect can be achieved with any spacer that sufficiently separates the Co-O planes. The question of water's particular role in the superconductivity is still very open, but we have shown that it has no effect on the electronic structure near the Fermi surface, other than to make it more two dimensional.

That our density functional-derived magnetic moments overestimate the observed moment of Na_xCoO_2 for all values of x, strongly implies that ferromagnetic quantum fluctuations are present in the system. We postulate that the superconductivity of the hydrated compound could arise through these fluctuations resulting in a new example of triplet state superconductivity.

It will be very interesting to see what experiment says about the symmetry of the superconducting state. If indeed it is a triplet, this material with its relatively high critical temperature will provide an excellent arena for testing theories of spin fluctuation mediated superconductivity.

Acknowledgments

We are grateful for helpful discussions with R. Asahi, G. Baskaran, H. Ding, T. Egami, M.Z. Hasan, W. Koshibae, D. Mandrus, I.I. Mazin, A.J. Millis, D.A. Papaconstantopoulos, W.E. Pickett, B.C. Sales, K. Sandeman, S.S. Saxena, A. J. Schofield and I. Terasaki. M.D.J. is supported by a National Research Council Associateship. Work at the Naval Research Laboratory is supported by the Office of Naval Research. Some computations were performed using facilities of the DoD HPCMO ASC and ARL centers. The DoD-AE code was used for some of this work.

References

[1] B. G. Levi, Physics Today, **15**, (2003)

[2] D. Carlier, A. Van der Ven, C. Delmas and G. Ceder, Chem. Mater. **15**, 2651 (2003); A. Van der Ven and G. Ceder, Electrochemical and Solid-State Letters **3**, Ven1 (2000), and refs. therein.

[3] Von M. Jansen and R. Hoppe, Z. Anorg. Allg. Chem. **408**, 104 (1974).

[4] I. Terasaki, Y. Sasago and U. Uchinokura, Phys. Rev. B **56**, 12685 (1997); I. Terasaki, Physica B **328**, 63 (2003).

[5] S. Li, R. Funahashi, I. Matsubara, K. Ueno and H. Yamada, J. Mater. Chem. **9**, 1659 (1999).

[6] A.C. Masset, C. Michel, A. Maignan, M. Hervieu, O. Toulemonde, F. Studer, B. Raveau and J. Hejtmanek, Phys. Rev. B **62**, 166 (2000).

[7] R. Funahashi, I. Matsubara, H. Ikuta, T. Takeuchi, U. Mizutani and S. Sodeoka, Jpn. J. Appl. Phys., Part 2 **39**, L1127 (2000).

[8] S. Hebert, S. Lambert, D. Pelloquin and A. Maignan, Phys. Rev. B **64**, 172101 (2001).

[9] Y. Miyazaki, M. Onoda, T. Oku, M. Kikuchi, Y. Ishii, Y. Ono, Y. Morii and T. Kajitani, J. Phys. Soc. Jpn. **71**, 491 (2002).

[10] T. Yamamoto, K. Uchinokura and I. Tsukada, Phys. Rev. B **65**, 184434 (2002).

[11] M. Hervieu, A. Maignan, C. Michel, V. Hardy, N. Creon and B. Raveau, Phys. Rev. B **67**, 045112 (2003).

[12] D. Pelloquin, A. Maignan, S. Hebert, C. Michel and B. Raveau, J. Solid State Chem. **170**, 374 (2003).

[13] D.J. Singh, Phys. Rev. B **61**, 13397 (2000).

[14] R. Asahi, J. Sugiyama and T. Tani, Phys. Rev. B **66**, 155103 (2002).

[15] D.J. Singh, Phys. Rev. B **68**, 020503 (2003).

[16] J. Kunes, K. -W. Lee, W. E. Pickett, cond-mat/0308388, (2003).

[17] G. Baskaran, Phys. Rev. Lett. **91**, 097003 (2003).

[18] Y. Ando, N. Miyamoto, K. Segawa, T. Kawata and I. Terasaki, Phys. Rev. B **60**, 10580 (1999).

[19] W. Koshibae and S. Maekawa, Phys. Rev. Lett. **87**, 236603 (2001).

[20] W. Koshibae and S. Maekawa, Physica B **329**, 896 (2003).

[21] Y.Y. Wang, N.S. Rogado, R.J. Cava and N.P. Ong, Nature **423**, 425 (2003).

[22] A. Maignan, S. Hebert, M. Hervieu, C. Michel, D. Pelloquin and D. Khomskii, J. Phys. Cond. Mat. **15**, 2711 (2003).

[23] I. Tsukada, T. Yamamoto, M. Takagi, T. Tsubone, S. Konno and K. Uchinokura, J. Phys. Soc. Jpn. **70**, 834 (2001).

[24] S.S. Saxena, P. Agarwal, K. Ahilan, F.M. Grosche, R.K.W. Haselwimmer, M.J. Steiner, E. Pugh, I.R. Walker, S.R. Julian, P. Monthoux, G.G. Lonzarich, A. Huxley, I. Sheikin, D. Braithwaite, and J. Flouquet, Nature **406**, 587 (2000).

[25] D. Aoki, A. Huxley, E. Ressouche, D. Braithwaite, J. Floquet, J.P. Brison, E. Lhotel and C. Paulsen, Nature **413**, 613 (2001).

[26] C. Pfleiderer, M. Uhlarz, S.M. Hayden, R. Vollmer, H. von Lohneysen, N.R. Bernhoeft, and G.G. Lonzarich, Nature **412**, 58 (2001).

[27] An alternate, Fulde-Ferrel-Larkin-Ovchinikov, non-triplet state is not excluded in $ZrZn_2$. [60]

[28] Y. Maeno, H. Hashimoto, K. Yoshida, S. Nishizaki, T. Fujita, J.G. Bednorz, and F. Lichtenberg, Nature **372**, 532 (1994).

[29] T.M. Rice and H. Sigrist, J. Phys. Condens. Matter **7**, L643 (1995).

[30] D. Fay and J. Appel, Phys. Rev. B **22**, 3173 (1980).

[31] P.B. Allen and B. Mitrovic, Solid State Phys. **37**, 1 (1982).

[32] K. Machida, and T. Ohmi, Phys. Rev. Lett. **86**, 850 (2001).

[33] D. Belitz and T.R. Kirkpatrick, Phys. Rev. Lett. **89**, 247202 (2002).

[34] T.R. Kirkpatrick and D. Belitz, Phys. Rev. B **67**, 024515 (2003).

[35] R. Roussev and A.J. Millis, Phys. Rev. B **63**, 140504 (2001).

[36] G. Santi, S.B. Dugdale and T. Jarlborg, Phys. Rev. Lett. **87**, 247004 (2001).

[37] I.I. Mazin and D.J. Singh, Phys. Rev. Lett. **82**, 4324 (1999).

[38] Y. Sidis, M. Braden, P. Bourges, B. Hennion, S. NishiZaki, Y. Maeno, and Y. Mori, Phys. Rev. Lett. **83**, 3320 (1999)

[39] A.T. Boothroyd, R. Coldea, D.A. Tennant, D. Prabhakaran, C.D. Frost, cond-mat/0312589

[40] I.I. Mazin and D.J. Singh, Phys. Rev. Lett. **79**, 733 (1997).

[41] K. Takada, H. Sakurai, E. Takayama-Muromachi, F. Izumi, R.A. Dilanian and T. Sasaki, Nature **422**, 53 (2003).

[42] R.E. Schaak, T. Klimczuk, M.L. Foo and R.J. Cava, Nature **424**, 527 (2003).

[43] H. Sakura, K. Takada, F. Izumi, D. A. Dilanian, R. Sasaki, E. Takayama-Muromachi, cond-mat/0310717.

[44] G. Baskaran, cond-mat/0306569.

[45] B. Lorenz, J. Cmaidalka, R. L. Meng Phys. Rev. B, **68**, 132504, (2003).

[46] M. L. Foo, R. E. Schaak, V. L. Miller, T. Klimczuk, N. S. Rogado, Y.Wang, G. C. Lau, C. Craley, H. W. Zandbergen, N. P. Ong, R. J. Cava, Solid State Commun. **127**, 33 (2003).

[47] C. Honerkamp, Phys. Rev. B **68**, 104510 (2003).

[48] M. Ogata, J. Phys. Soc. Japan, **72**, 1839 (2003).

[49] S. Park, Y. Lee, A. Moodenbaugh, T. Vogt, Phys. Rev. B **68** 180505, (2003).

[50] P. Blaha, K. Schwarz, G. K. H. Madsen, D. Kvasnicka, J. Luitz, WIEN2K, An Augmented Plane Wave + Local Orbitals Program for Calculating Crystal Properties (Karlheinz Schwarz, Techn. Universitat Wien, Austria), ISBN 3-9501031-1-2, 2201, J. Phys. Chem. Sol. **63**, (2002).

[51] D.J. Singh, Planewaves, Pseudopotentials and the LAPW Method (Kluwer Academic, Boston, 1994).

[52] J. W. Lynn, Q. Huang, C. M. Brown, V. L. Miller, M. L. Foo, R. E. Schaak, C. Y. Jones, E. A. Mackey, R. J. Cava, cond-mat/0307623.

[53] J. D. Jorgensen, M. Avdeev, D. G. Hinks, J. C. Burley, S. Short, cond-mat/0307627.

[54] R. Jin, B. C. Sales, P. Khalifah, D. Mandrus, Phys. Rev. Lett., **91** 217001, (2003).

[55] T. Egami, private communication

[56] D. Singh, Phys. Rev. B **43**, 6388 (1991).

[57] D.J. Singh, Phys. Rev. B **61**, 13397 (2000).

[58] Note that this is not really a constant factor. At lower hole concentrations the antiferromagnetic instability is expected to disappear before the ferromagnetic one, likely leading to a smaller and smaller region of the zone where magnetic instabilities exist. This would be consistent with ordering appearing as generically one expects the strength of the quantum fluctuations to be small when the region in **k** of magnetic instabilities is small. For lower Na concentration (higher p) the antiferromagnetism may become dominant as in Hubbard models approaching integer electron counts.

[59] A. Aguayo and D.J. Singh, Phys. Rev. B **66**, 020401 (2002).

[60] D.J. Singh and I.I. Mazin, Phys. Rev. Lett. **88**, 187004 (2002).

[61] D.J. Singh and I.I. Mazin, Phys. Rev. B **63**, 165101 (2001).

[62] S.A. Grigera, R.S. Perry, A.J. Schofield, M. Chiao, S.R. Julian, G.G. Lonzarich, S.I. Ikeda, Y. Maeno, A.J. Millis, and A.P. Mackenzie, Science **294**, 329 (2001).

[63] H. Yamada, K. Fukamichi and T. Goto, Phys. Rev. B **65**, 024413 (2001).

[64] A.J. Millis, A.J. Schofield, G.G. Lonzarich and S.A. Grigera, Phys. Rev. Lett. **88**, 217204 (2002).

[65] M. Shimizu, Rep. Prog. Phys. **44**, 329 (1981).

[66] P. Larson, I.I. Mazin and D.J. Singh, cond-mat/0305407.

HOLSTEIN-PRIMAKOFF REPRESENTATION FOR STRONGLY CORRELATED ELECTRON SYSTEMS

Siyavush Azakov

Institute of Physics, Azerbaijan Academy of Sciences, Baku, Azerbaijan

Abstract First we show that the algebra of operators entering the Hamiltonian of the $t-J$ model describing the strongly correlated electron system is graded $spl(2.1)$ algebra. Then after a brief discussion of its atypical representations we construct the Holstein-Primakoff nonlinear realization of these operators which allows to carry out the systematic semiclassical approximation, similarly to the spin-wave theory of localized magnetism. The fact that the $t-J$ model describes the itinerant magnetism is reflected in the presence of the spinless fermions.

For the supersymmetric $spl(2.1)$ algebra the supercoherent states are proposed and the partition function of the $t-J$ model is represented as a path integral with the help of these states.

Keywords: Supersymmetry, itinerant magnetism, spin-wave theory, supercoherent states.

7.1 Introduction

The discovery of high transition temperature ceramic superconductors has renewed interest in the study of strongly correlated electron systems, since it is widely believed that the anomalous properties of such materials are related to the strong Coulomb repulsion of electrons [1].

One of the most interesting models which has received much attention is the two-dimensional single-band Hubbard model [2]. Its Hamiltonian describes a single electron band in a tight binding basis, with an on-site electron-electron repulsion for electrons of opposite spin.

$$\hat{H} = -t \sum_{\langle i,j \rangle, \sigma=\uparrow,\downarrow} (\hat{c}^{\dagger}_{i\sigma}\hat{c}_{j\sigma} + \hat{c}^{\dagger}_{j\sigma}\hat{c}_{i\sigma}) + U \sum_{i} \hat{n}_{i\uparrow}\hat{n}_{i\downarrow} \ , \tag{1}$$

where $\hat{n}_{i\sigma} = \hat{c}^{\dagger}_{i\sigma}\hat{c}_{i\sigma}$ is the operator of a number of electrons at the site i with spin projection σ, the symbol $\langle i, j \rangle$ indicates ordered $(i < j)$

S. Halilov (ed.), Physics of Spin in Solids: Materials, Methods and Applications, 101–114.

nearest-neighbor pairs, $U > 0$. At each site of the lattice we have four states

$$|0\rangle\ ,\ |\sigma\rangle = \hat{c}^{\dagger}_{\sigma}|0\rangle\ ,\ \sigma = \uparrow (1), \downarrow (-1)\ ,\ |2\rangle = |\uparrow\downarrow\rangle = \hat{c}^{\dagger}_{\uparrow}\hat{c}^{\dagger}_{\downarrow}|0\rangle\ , \qquad (2)$$

where $|0\rangle$ describes the empty lattice site, which serves as the vacuum for electron operators, i.e. $\hat{c}_{\sigma}|0\rangle = 0$, $|\sigma\rangle$ is the state of the singly occupied site with the spin projection σ, and $|2\rangle$ is the state describing the doubly occupied site.

The Hubbard Hamiltonian (1) may be rewritten in terms of the Hubbard operators which are the projection operators. For each site i the Hubbard operators are

$$\hat{X}^{AB}_i = |A\rangle_{ii}\langle B|\ , \qquad (3)$$

where $|A\rangle_i = (|0\rangle_i,\ |\sigma\rangle_i,\ |2\rangle_i)$, and

$$\hat{H} = -t \sum_{(i,j),\sigma} \left(\hat{X}^{\sigma 0}_i + \sigma \hat{X}^{2\bar{\sigma}}_i\right)\left(\hat{X}^{0\sigma}_j + \sigma \hat{X}^{\bar{\sigma}2}_j\right) + U \sum_i \hat{X}^{22}_i \qquad (4)$$

(i,j) denotes a nearest neighbor pairs, $\bar{\sigma} = -\sigma$. The Hubbard operators obey a simple multiplication law $\hat{X}^{AB}_i \hat{X}^{CD}_i = \delta_{BC}\hat{X}^{AD}_i$ and the following supercommutation relations

$$\left[\hat{X}^{AB}_i, \hat{X}^{CD}_j\right]_s = \delta_{ij}\left(\hat{X}^{AD}_i \delta_{BC} - (-1)^{\chi^{AB}_i \chi^{CD}_j} \hat{X}^{CB}_i \delta_{AD}\right), \qquad (5)$$

where $s = 2\left\{\theta\left(\chi^{AB} + \chi^{CD} - 3/2\right) - 1/2\right\}$ and the graded characters χ^{AB} of the Hubbard operators are 0 in the case of the bosonic operators $\left(\hat{X}^{00}, \hat{X}^{22}, \hat{X}^{\sigma\sigma}, \hat{X}^{\sigma\bar{\sigma}}, \hat{X}^{02}, \hat{X}^{20}\right)$ and 1 in the case of the fermionic operators $\left(\hat{X}^{0\sigma}, \hat{X}^{\sigma 0}, \hat{X}^{2\sigma}, \hat{X}^{\sigma 2}\right)$. Supercommutation relations (5) show that 16 Hubbard operators $\{X^{AB}\}$ are generators of the $u(2.1)$ superalgebra and Eqn.(3) fixes its representation.

For superconductivity the most relevant are the low-energy excitations in the strong coupling regime, where $U \gg t$.

Doing the strong coupling expansion (in parameter t/U) in the second order one gets an effective Hamiltonian which acts on a Hilbert space where states with doubly occupied sites are excluded [3]. This Hamiltonian has the same low-energy spectrum as the original Hamiltonian. It is called the $t - J$ model Hamiltonian [6] and has the following form

$$\hat{H}_{t-J} = -t \sum_{(i,j),a=1,2} \hat{X}^a_i \hat{X}_{aj} + \frac{J}{2} \sum_{(i,j)} \left(\hat{\mathbf{S}}_i \hat{\mathbf{S}}_j - \frac{1}{4}\hat{n}_i \hat{n}_j\right), \qquad (6)$$

where $J = \frac{4t^2}{U}$ and $\hat{X}_a, \hat{\mathbf{S}}, \hat{n}$ denote the following combinations of electron operators (to simplify the notations we drop the site index). Hole operators:

$$\hat{X}_1 \equiv \hat{X}^{0\uparrow} = (1-\hat{n}_\downarrow)\hat{c}_\uparrow \quad , \quad \hat{X}_2 \equiv \hat{X}^{0\downarrow} = (1-\hat{n}_\uparrow)\hat{c}_\downarrow \quad , \hat{X}^a \equiv \hat{X}_a^\dagger \ . \quad (7)$$

Spin operators:

$$\hat{\mathbf{S}} = \frac{1}{2}\sum_{\sigma,\sigma'} \hat{c}_\sigma^\dagger \hat{\boldsymbol{\sigma}}_{\sigma\sigma'}\hat{c}_{\sigma'} = \left(\hat{S}^x, \hat{S}^y, \hat{S}^z\right) \equiv \left(\hat{S}_1, \hat{S}_2, \hat{S}_3\right) , \quad (8a)$$

($\hat{\sigma}^x, \hat{\sigma}^y, \hat{\sigma}^z$ are the usual Pauli matrices)

$$\hat{S}^+ = \hat{S}^x + i\hat{S}^y = \hat{X}^{\uparrow\downarrow} \quad , \quad \hat{S}^- = \left(\hat{S}^+\right)^\dagger = \hat{X}^{\downarrow\uparrow} \quad , \quad \hat{S}^z = \frac{1}{2}\left(\hat{X}^{\uparrow\uparrow} - \hat{X}^{\downarrow\downarrow}\right) . \quad (8b)$$

Charge operator:

$$\hat{n} = \sum_\sigma \hat{n}_\sigma = 1 - \hat{X}^{00} . \quad (9)$$

The set of these operators operating in the restricted Hilbert space form a representation of the graded (supersymmetric) Lie algebra $spl(2.1)$, which is the graded extension of the Lie algebra $su(2)$. In fact, introducing the operator

$$\hat{S}_0 = 1 - \frac{1}{2}\hat{n} \quad (10)$$

one can easily check that the operators (7),(8) and (10) satisfy the following commutation/anti-commutation relations

$$\begin{aligned}
[\hat{S}_\mu, \hat{S}_\nu] &= i\varepsilon^\lambda_{0\mu\nu}\hat{S}_\lambda \ , \quad \mu,\nu,\lambda = 0,1,2,3 \ , && (11a)\\
[\hat{X}^a, \hat{S}_\mu] &= \frac{1}{2}(\sigma^\mu)^a_b \hat{X}^b \ , \quad a,b = 1,2 \ , && (11b)\\
\{\hat{X}_a, \hat{X}^b\} &= (\sigma^\mu)^b_a \hat{S}_\mu \ , && (11c)\\
\{\hat{X}_a, \hat{X}_b\} &= 0 \ . && (11d)
\end{aligned}$$

The summation over repeated "Lorentz" and 'spinorial" indices is assumed, $\sigma_0 = \mathbf{1}_2, \hat{S}_\mu = (\hat{S}_0, \hat{\mathbf{S}}), \varepsilon^3_{012} = 1, (\sigma^\mu)^b_a = (\delta_{ab}, \boldsymbol{\sigma}_{ab})$. The greek four-vector indices are raised and lowered using the metric tensor $g_{\mu\nu} = diag(1,-1,-1,-1)$. Note that the even sector of the algebra is $su(2) \times u(1)$, where $su(2)$ corresponds to the spin degrees of freedom and $u(1)$ to the charge degree of freedom. The generator $\hat{S}_0$ should be introduced in order to close the $spl(2.1)$ algebraic rules (11b) and (11c). We must specify the commutation rules between operators defined at different sites as well. We declare that bosonic operator defined at a given site always commutes with an operator defined at another site, while fermionic operators defined at two different sites always anticommute.

Thus, the $t-J$ model which describes a system of itinerant magnetism has a Hamiltonian which is bilinear in a set of operators belonging to a graded Lie algebra, whereas the Hamiltonian of the Heisenberg model of antiferromagnet to which the $t-J$ model reduces at half-filling (when at each site we have only one electron) is bilinear in a set of operators belonging to a usual Lie algebra. Hence, as suggested by Wiegmann [7], models of itinerant magnetism can be considered as a supersymmetric extension of models of localized magnetism: the hole being the superpartner of the spin.

The operators which enter the Hamiltonian (6) of the $t-J$ model belong to the fundamental representation of the $spl(2.1)$ algebra. In the restricted Hilbert space where doubly occupied sites are excluded and states are $|0\rangle, |\uparrow\rangle$ and $|\downarrow\rangle$ they are represented by 3×3 matrices.

Our aim now is to consider them in an arbitrary representation of the algebra. This is the standard way to develop the spin-wave technique for quantum spin systems, by introducing Holstein-Primakoff (HP) representation [8]. It is worthwhile to note, that when we speak about a specific model we assume a specific choice of the representation for the operators. The dynamics described by the model and the physical consequences usually depend on the chosen representation.

Spin-wave theory helps us considerably in the understanding of the spin $\frac{1}{2}$ quantum ferromagnetic or antiferromagnetic Heisenberg model. HP representation allows to develop systematic semiclassical approximation which is $1/s$ expansion (where s is the eigenvalue of the spin operator). We would like to mention that the most accurate Monte-Carlo simulations done for 2-dimensional Heisenberg antiferromagnet with spin 1/2 not so long ago [9] gave the values of the ground state energy and the sublattice magnetization which differ less than 1/1000 with those obtained by the second order spin-wave theory in 1960's [10]. Up to now nobody can theoretically explain this striking agreement. Probably the large-s limit captures the essential physics, since the generalization to higher spin preserves the symmetries of the original model.

HP representation is also effectively used in the reduction of the model of Heisenberg antiferromagnet to the nonlinear σ-model [11]. This reduction plays an essential role in our understanding of the one-dimensional Heisenberg antiferromagnet where charge and spin excitations are separated and the ground state is not a Fermi liquid. It has been suggested that this spin-charge separation may also occur in two dimensions and is responsible for the unusual normal state properties found in the cuprate superconductors.

The rest of the paper is organized as follows. Sec.2 is devoted to a brief review of the irreducible representations of the *spl*(2.1) algebra, and more details of the so-called atypical representations which are relevant for the strongly correlated electron system are given. In Sec.3 we introduce slave particles and define HP representation for these operators considered in atypical representations. In the Hamiltonian approach HP representation allows to obtain systematically the leading order of the semiclassical approximation and the corrections. In Sec.4 we construct coherent states for the *spl*(2.1) algebra, which is a graded algebra, so they can be called supercoherent states. Using these states one can obtain a partition function of the $t-J$ model as a path integral in the form which again can be used to develop a systematic semiclassical approximation.

The final section concludes with a discussion of the results and possible future development.

7.2 Representations of the *spl*(2.1) algebra

The representation theory for the *spl*(2.1) algebra has been studied in detail in [12], and several classes of representations were found. It was shown in this paper that the states of a finite dimensional irreducible representation are labeled by the eigenvalues of the operators $\hat{S}_0, \hat{\mathbf{S}}^2$ and $\hat{S}^z$ which we denote by $s_0, s(s+1)$ and m respectively. A general irreducible representation has an arbitrary complex s_0, and integer or half-integer s. As it is shown in [12] there are at most four multiplets

$$|s_0, s, m\rangle \ , \qquad m = -s, -s+1, \ldots, s \tag{12a}$$
$$|s_0 + \tfrac{1}{2}, s - \tfrac{1}{2}, m'\rangle \ , \qquad m' = -s + \tfrac{1}{2} \ldots, s - \tfrac{1}{2} \tag{12b}$$
$$|s_0 - \tfrac{1}{2}, s - \tfrac{1}{2}, m'\rangle \ , \qquad m' = -s + \tfrac{1}{2}, \ldots, s - \tfrac{1}{2} \tag{12c}$$
$$|s_0, s-1, m\rangle \ , \qquad m = -s+1, \ldots, s-1 \ , \tag{12d}$$

which transform among themselves under the action of operators (7)-(10). The multiplets (12a) and (12d) are called even and the multiplets (12b) and (12c) are called odd.

The even operators acting on these states preserve the parity

$$\hat{S}_0|s_0, s, m\rangle = s_0|s_0, s, m\rangle \ , \tag{13a}$$
$$\hat{S}^z|s_0, s, m\rangle = m|s_0, s, m\rangle \ , \tag{13b}$$
$$\hat{S}^{\pm}|s_0, s, m\rangle = \sqrt{(s \mp m)(s \pm m + 1)}|s_0, s, m \pm 1\rangle \ . \tag{13c}$$

The same relations hold for (12b)-(12c). Instead the odd operators change the parity of the state e.g.:

$$\hat{X}_a|s_0, s, m\rangle$$
$$= e^{i\alpha}\sqrt{s-(-1)^a m}\;|s_0+\tfrac{1}{2}, s-\tfrac{1}{2}, m+(-1)^a\tfrac{1}{2}\rangle \qquad (14a)$$
$$\hat{X}^a|s_0, s, m\rangle = 0 \qquad (14b)$$
$$\hat{X}_a|s_0+\tfrac{1}{2}, s-\tfrac{1}{2}, m'\rangle = 0 \qquad (14c)$$
$$\hat{X}^a|s_0+\tfrac{1}{2}, s-\tfrac{1}{2}, m'\rangle$$
$$= e^{-i\alpha}\sqrt{s-(-1)^a m'+\tfrac{1}{2}}\;|s_0, s, m'-(-1)^a\tfrac{1}{2}\rangle\;, \qquad (14d)$$

The phase α should be introduced since the relative normalization of the even and odd multiplets is not fixed apriori: however, different choices of α lead to equivalent irreducible representations. For convenience we set $\alpha = 0$.

The algebra $spl(2.1)$ is a rank-2 graded Lie algebra and has just two Casimir operators [12]

$$\hat{K}_2 = -\hat{S}^\mu\hat{S}_\mu + \frac{1}{2}\hat{X}_a\hat{X}^a - \frac{1}{2}\hat{X}^a\hat{X}_a\;, \qquad (15a)$$
$$\begin{aligned}\hat{K}_3 &= \hat{S}_0(\hat{X}_a\hat{X}^a - \hat{X}^a\hat{X}_a - \hat{S}^\mu\hat{S}_\mu)\\ &+ \frac{1}{6}[\epsilon^{ab}\hat{X}_a(\hat{\mathbf{S}}\cdot\boldsymbol{\sigma})_{bc}\hat{X}^c + \hat{X}^a(\hat{\mathbf{S}}\cdot\boldsymbol{\sigma})_{ab}\epsilon^{bc}\hat{X}_c]\\ &+ \frac{1}{12}[\epsilon^{ab}\hat{X}_a\hat{X}^c(\hat{\mathbf{S}}\cdot\boldsymbol{\sigma})_{bc} + \hat{X}^a\epsilon^{bc}\hat{X}_c(\hat{\mathbf{S}}\cdot\boldsymbol{\sigma})_{ab}]\;,\end{aligned} \qquad (15b)$$

where ϵ^{ab} is the antisymmetric tensor ($\epsilon^{12} = 1$). The Casimir operators (15) have eigenvalues which look simply in terms of s_0 and s ; $K_2 = s^2 - s_0^2$, $K_3 = s_0(s^2 - s_0^2)$. There are two classes of representations which are called *atypical* and those for which K_2 and K_3 are equal to zero, i.e. those where $s_0 = s$ and $s_0 = -s$. These representations are called atypical because for them in contrast with the usual Lie algebras, the eigenvalues of the Casimir operators do not specify the irreducible representations.

The $s_0 = s$ and $s_0 = -s$ atypical irreducible representations are isomorphic. The states of the former are (12a) and (12b) and of the latter are (12c) and (12d), and in both cases the dimensionality of the irreducible representation is $4s + 1$. For the physical systems we wish to study the relevant representation is the $s_0 = s$ atypical irreducible representation, and, in the future, we will deal only with two multiplets (12a) and (12b) in the case when $s_0 = s$. By choosing the generic state vector in column-vector form

$$|\Psi\rangle = \begin{pmatrix} |s+\frac{1}{2}, s-\frac{1}{2}, m'\rangle \\ |s, s, m\rangle \end{pmatrix} \tag{16}$$

we can present the generators $\hat{S}_\mu, \hat{X}_a, \hat{X}^a$ in the matrix form. For the even generators we have the block diagonal form

$$\hat{S}_0 = \begin{pmatrix} (s+\frac{1}{2})\mathbf{1}_{2s} & 0 \\ 0 & s\mathbf{1}_{2s+1} \end{pmatrix}, \tag{17a}$$

$$\hat{\mathbf{S}} = \begin{pmatrix} \hat{\mathbf{S}}_{2s} & 0 \\ 0 & \hat{\mathbf{S}}_{2s+1} \end{pmatrix}, \tag{17b}$$

where $\hat{\mathbf{S}}_n$ are the $su(2)$ algebra generators in the n-dimensional representation, and $\mathbf{1}_n$ is the $n \times n$ identity matrix. The odd generators have the block-off-diagonal form

$$\hat{X}_a = \begin{pmatrix} 0 & D_a \\ 0 & 0 \end{pmatrix}, \quad \hat{X}^a = \begin{pmatrix} 0 & 0 \\ D_a^\dagger & 0 \end{pmatrix}, \tag{18}$$

where the rectangular $(2s+1) \times 2s$ D_1 and D_2 matrices are defined as

$$D_1 = \begin{pmatrix} \sqrt{2s} & \cdots & 0 & 0 & 0 \\ \vdots & & \vdots & \vdots & \vdots \\ 0 & \cdots & \sqrt{2} & 0 & 0 \\ 0 & \cdots & 0 & \sqrt{1} & 0 \end{pmatrix}, \tag{19a}$$

$$D_2 = \begin{pmatrix} 0 & \sqrt{1} & 0 & \cdots & 0 \\ 0 & 0 & \sqrt{2} & \cdots & 0 \\ \vdots & \vdots & \vdots & \vdots & \vdots \\ 0 & 0 & 0 & \cdots & \sqrt{2s} \end{pmatrix}. \tag{19b}$$

The Hubbard operators (7) and (8) belong to the fundamental $s = \frac{1}{2}$ atypical irreducible representation of $spl(2.1)$, and hence the single-particle physical states of strongly correlated electron system carry this representation.

The odd generators of $spl(2.1)$ enter the hopping terms of the Hamiltonian (6), which can be thought of as the operators which destroy or create a hole in favor of a spin or vice versa. Eqs. (14) explicitly expose this property: the odd generators are step operators which interchange states with different parity by raising or lowering by a half unit the eigenvalues s_0 and s.

The matrices (17) and (18) are the higher-dimensional generalizations of the Hubbard matrices and the states (12a) and (12b) (where $s_0 = s$) are the generalized "spin" and "hole" state, respectively, on which the above mentioned operators act. One can introduce the electric charge operator $\hat{\mathcal{E}} = 2\hat{S}_0 - 2s\hat{I}$, so that the states (12a) have $\mathcal{E} = 0$ and the states (12b) have $\mathcal{E} = 1$. The conservation of the total electron number can be restated as the conservation of the total charge if we consider the original $s = \frac{1}{2}$ representation

$$\sum_i \hat{\mathcal{E}}_i = \sum_i (2\hat{S}_{0i} - 2s\hat{I}) = N_h \equiv M - N_{el}, \tag{20}$$

where N_h is the number of holes, M is the number of sites and N_{el} is the number of electrons.

7.3 Slave particles. Holstein-Primakoff representation

Now we may proceed in the same way as we usually do for the spin operators [4] . We introduce at each site i two Bose operators $\hat{b}_{1i}, \hat{b}_{2i}$ and one spinless Fermi operator $\hat{f}_i$

$$\begin{aligned} [\hat{b}_{i\alpha}, \hat{b}^\dagger_{j\beta}] = \delta_{\alpha\beta}\delta_{ij}, \quad , \quad [\hat{b}_{i\alpha}, \hat{b}_{j\beta}] = [\hat{b}^\dagger_{i\alpha}, \hat{b}^\dagger_{j\beta}] = 0 \ , \\ \{\hat{f}_i, \hat{f}^\dagger_j\} = \delta_{ij} \quad , \quad \{\hat{f}_i, \hat{f}_j\} = \{\hat{f}^\dagger_i, \hat{f}^\dagger_j\} = 0 \ , \end{aligned} \tag{21}$$

which obey the holonomic constraint at each site

$$\sum_{\alpha=1}^{2} \hat{b}^\dagger_{i\alpha}\hat{b}_{i\alpha} + \hat{f}^\dagger_i \hat{f}_i = 2s \cdot \mathbf{1} \ . \tag{22}$$

Then one can easily check that the algebra (11) of operators $\hat{S}_\mu, \hat{X}_a$, is satisfied if we choose the following representation (again we drop the site index)

$$\begin{aligned} \hat{\mathbf{S}} = \frac{1}{2} \sum_{\alpha,\beta=1,2} \hat{b}^\dagger_\alpha \boldsymbol{\sigma}_{\alpha\beta}\hat{b}_\beta \quad , \quad \hat{S}_0 = \mathbf{1}s + \tfrac{1}{2}\hat{f}^\dagger\hat{f} \quad , \\ \hat{X}_1 = \hat{f}^\dagger\hat{b}_1 \quad , \quad \hat{X}_2 = \hat{f}^\dagger\hat{b}_2 \end{aligned} \tag{23}$$

and the $(4s+1)$ states (12a) and (12b) are represented by

$$\begin{aligned} |s,s,m\rangle = & \frac{1}{\sqrt{(s+m)!(s-m)!}} \\ & \times (\hat{b}^\dagger_1)^{s+m}(\hat{b}^\dagger_2)^{s-m}|0\rangle_b|0\rangle_f \ , \end{aligned} \tag{24a}$$

$$|s+\tfrac{1}{2}, s-\tfrac{1}{2}, m'\rangle = \frac{e^{i\Lambda}}{\sqrt{(s-\frac{1}{2}+m')!(s-\frac{1}{2}-m')!}}$$
$$\times(\hat{b}_1^\dagger)^{s-\frac{1}{2}+m'}(\hat{b}_2^\dagger)^{s-\frac{1}{2}-m'}\hat{f}^\dagger|0\rangle_b|0\rangle_f \ , \tag{24b}$$

where Λ is a phase, which should be present because the relative normalization of the multiplets (12a) and (12b) is not apriori fixed. $|0\rangle_b(|0\rangle_f)$ is the vacuum for $\hat{b}_\alpha(\hat{f})$ operators [5].

In order to get the HP representation we will follow the procedure used in the pure spin case [4]. Then the $(4s+1)$ dimensional Hilbert space spanned by the states (12a), (12b) is put into correspondence with the Hilbert space generated by the states $(\hat{b}_2^\dagger)^n|s,s,s\rangle$ and $(\hat{b}_2^\dagger)^n|s+\frac{1}{2}, s-\frac{1}{2}, s-\frac{1}{2}\rangle$ satisfying the relation $\hat{b}_2^\dagger\hat{b}_2 + \hat{f}^\dagger\hat{f} \le 2s$. In this Hilbert space the $\hat{b}_1$ operator can be excluded with the help of the constraint (22)

$$\hat{b}_1 = \sqrt{2s - \hat{b}^\dagger\hat{b} - \hat{f}^\dagger\hat{f}}, \tag{25}$$

where $\hat{b} \equiv \hat{b}_2$ and it can be considered as a Fock space with the vacuum $|\Phi_0\rangle = |s,s,s\rangle$ and for each state we have $\hat{b}^\dagger\hat{b} + \hat{f}^\dagger\hat{f} \le 2s$. In this space the operators can be represented as follows:

$$\begin{aligned}
&\hat{S}_0 = s + \tfrac{1}{2}\hat{f}^\dagger\hat{f} \ , \qquad \hat{X}_1 = \hat{f}^\dagger\sqrt{2s - \hat{b}^\dagger\hat{b} - \hat{f}^\dagger\hat{f}} \\
&\hat{S}^z = s - \hat{b}^\dagger\hat{b} - \tfrac{1}{2}\hat{f}^\dagger\hat{f} \ , \qquad \hat{X}_2 = \hat{f}^\dagger\hat{b} \\
&\hat{S}^+ = \sqrt{2s - \hat{b}^\dagger\hat{b} - \hat{f}^\dagger\hat{f}}\hat{b} \ , \qquad \hat{X}^1 = \sqrt{2s - \hat{b}^\dagger\hat{b} - \hat{f}^\dagger\hat{f}}\hat{f} \\
&\hat{S}^- = \hat{b}^\dagger\sqrt{2s - \hat{b}^\dagger\hat{b} - \hat{f}^\dagger\hat{f}} \ , \qquad \hat{X}^2 = \hat{b}^\dagger\hat{f} \ .
\end{aligned} \tag{26}$$

It is straightforward to check that the (single-site) commutation relations (11) are fulfilled and the two Casimir operators (15) are identically zero in this realization.

This is the generalization to $spl(2.1)$ of the usual Holstein-Primakoff representation for $su(2)$ spin algebra and it can be called a graded HP representation. For $s = \frac{1}{2}$ we have the following correspondence with the states of the restricted Hilbert space which we had from the very beginning: $|0\rangle = \hat{f}^\dagger|\Phi_0\rangle, |\uparrow\rangle = |\Phi_0\rangle, |\downarrow\rangle = \hat{b}^\dagger|\Phi_0\rangle$. We see that the $\hat{b}$ operator is a spin-flip operator.

In the HP representation the conservation of the total charge (20) takes the form

$$\sum_{i=1}^{M} \hat{f}_i^\dagger\hat{f}_i = N_h. \tag{27}$$

At half-filling (when holes are absent) the graded HP representation reduces to the standard one, and now it is explicitly seen that for this particular filling the $t - J$ model reduces to the Heisenberg model.

From the representation (26) we may refer to the semiclassical regime as the regime in which

$$\frac{1}{2s}\langle \hat{b}_i^\dagger \hat{b}_i + \hat{f}_i^\dagger \hat{f}_i \rangle \leq 1 \tag{28}$$

namely, when the spin "flip" or hole "flow" rate on each site is expected to be very small.

Whenever Eq.(28) is reliable, the square roots in Eqs.(26) can be expanded in powers of $1/s$. This approach generalizes the spin wave theory for the models describing localized magnetism to the models describing itinerant magnetism where the operators entering the Hamiltonian belong to the graded algebra, and it leads to a description of the system in terms of interacting bosons and fermions. The itinerant nature of the magnetism described by the $t - J$ model is seen in the presence of spinless fermions.

7.4 Supercoherent States for *spl*(2.1) Superalgebra

In order to construct coherent states in the graded case of the $spl(2.1)$ superalgebra we can proceed almost in the same way as we usually do in the case of the $su(2)$ algebra describing spin operators [13], since the even sector of $spl(2.1)$ is isomorphic with $su(2) \times u(1)$ algebra.

The coherent states will be constructed again in the $s_0 = s$ atypical representation with state vectors (16). One can show that the state $|\mathbf{N}, \eta\rangle$, which is described by the unit vector $\mathbf{N} = (\sin\vartheta\cos\varphi, \sin\vartheta\sin\varphi, \cos\vartheta)$ and the Grassmann variable η of the form

$$\begin{aligned} |\mathbf{N}, \eta\rangle &= \frac{e^{-\zeta^* \hat{S}^-}}{(1+|\zeta|^2)^s}|s, s, s\rangle + \eta \frac{e^{-\zeta^* \hat{S}^-}}{(1+|\zeta|^2)^{s-\frac{1}{2}}}|s+\tfrac{1}{2}, s-\tfrac{1}{2}, s-\tfrac{1}{2}\rangle \\ &= |\mathbf{N}\rangle_s |0\rangle_f + \eta |\mathbf{N}\rangle_{s-\frac{1}{2}} \hat{f}^\dagger |0\rangle_f \ , \end{aligned} \tag{29}$$

where $\zeta = -\tan\frac{\vartheta}{2} e^{-i\varphi}$ and $|\mathbf{N}\rangle_s$ is the spin coherent state in the space with spin s, has all the properties of the coherent state. Namely

$$\langle \mathbf{N}, \eta | \hat{S}_\mu | \mathbf{N}, \eta \rangle = (s, s\mathbf{N}) + \bar{\eta}\eta(s + \tfrac{1}{2}, (s - \tfrac{1}{2})\mathbf{N}) \ . \tag{30}$$

(we put $\Lambda = 0$). Using commutation relations

$$[\hat{X}_1, e^{\xi \hat{S}^-}] = 0, \qquad [\hat{X}_2, e^{\xi \hat{S}^-}] = \xi e^{\xi \hat{S}^-} \hat{X}_1 \ , \tag{31a}$$

$$[\hat{X}^2, e^{\xi \hat{S}^-}] = 0, \qquad [\hat{X}^1, e^{\xi \hat{S}^-}] = -\xi e^{\xi \hat{S}^-} \hat{X}^2 \ , \tag{31b}$$

we also get

$$\hat{X}_1 |\mathbf{N}, \eta\rangle = \sqrt{2s} \cos(\vartheta/2) |\mathbf{N}\rangle_{s-\frac{1}{2}} \hat{f}^\dagger |0\rangle_f \ , \tag{32a}$$

$$\hat{X}_2 |\mathbf{N}, \eta\rangle = \sqrt{2s} e^{i\varphi} \sin(\vartheta/2) |\mathbf{N}\rangle_{s-\frac{1}{2}} \hat{f}^\dagger |0\rangle_f \ , \tag{32b}$$

$$\hat{X}^1 |\mathbf{N}, \eta\rangle = \frac{\eta}{\cos(\vartheta/2)} \left(\sin(\vartheta/2) e^{i\varphi} \hat{b}^\dagger |\mathbf{N}\rangle_{s-\frac{1}{2}} - \sqrt{2s} |\mathbf{N}\rangle_{\mathbf{s}} \right) |0\rangle_f \tag{33a}$$

$$\hat{X}^2 |\mathbf{N}, \eta\rangle = -\eta \hat{b}^\dagger |\mathbf{N}\rangle_{s-\frac{1}{2}} |0\rangle_f \ , \tag{33b}$$

and

$$\begin{aligned} \langle \mathbf{N}, \eta | \hat{X}_1 | \mathbf{N}, \eta \rangle &= \sqrt{2s} \cos(\vartheta/2) \bar{\eta}, \\ \langle \mathbf{N}, \eta | \hat{X}_2 | \mathbf{N}, \eta \rangle &= \sqrt{2s} \sin(\vartheta/2) e^{i\varphi} \bar{\eta} \ , \end{aligned} \tag{34a}$$

$$\begin{aligned} \langle \mathbf{N}, \eta | \hat{X}^1 | \mathbf{N}, \eta \rangle &= \sqrt{2s} \cos(\vartheta/2) \eta, \\ \langle \mathbf{N}, \eta | \hat{X}^2 | \mathbf{N}, \eta \rangle &= \sqrt{2s} \sin(\vartheta/2) e^{-i\varphi} \eta. \end{aligned} \tag{34b}$$

The overlap of two supercoherent states takes the form

$$\langle \mathbf{N}_1, \eta_1 | \mathbf{N}_2, \eta_2 \rangle = {}_s\langle \mathbf{N}_1 | \mathbf{N}_2 \rangle_s \left[1 + \bar{\eta}_1 \eta_2 \left({}_{\frac{1}{2}}\langle \mathbf{N}_1 | \mathbf{N}_2 \rangle_{\frac{1}{2}} \right)^{-1} \right] \tag{35}$$

and

$${}_s\langle \mathbf{N}_1 | \mathbf{N}_2 \rangle_s = e^{i\Phi(\mathbf{N}_1, \mathbf{N}_2, \mathbf{N}_0)s} \left(\frac{1 + \mathbf{N}_1 \cdot \mathbf{N}_2}{2} \right)^s . \tag{36}$$

We have the resolution of unity (in the pure spin case):

$$\int d\mu_s(\mathbf{N}) |\mathbf{N}\rangle_{ss}\langle \mathbf{N}| = \mathbf{1} \tag{37}$$

where

$$\int d\mu_s(\mathbf{N}) \ldots = \frac{2s+1}{4\pi} \int_0^\pi \sin\vartheta d\vartheta \int_0^{2\pi} d\varphi \ldots \ . \tag{38}$$

For the supercoherent states the resolution of unity will take the form

$$\int d\mu_{s-\frac{1}{2}}(\mathbf{N}) d\bar{\eta} d\eta e^{-\gamma_s \bar{\eta}\eta} |\mathbf{N}, \eta\rangle\langle \mathbf{N}, \eta| = \mathbf{1} \ , \tag{39}$$

where $\gamma_s = \frac{2s+1}{2s}$.

We can now construct a path integral representation for the partition function almost in the same manner as we do for pure spin systems [14], [4].

For arbitrary spin we will generalize (10) as follows

$$\hat{n}_i = -2\hat{S}_{0i} + (2s+1)\mathbf{1}. \tag{40}$$

So from (6),(30) and (34) we get

$$\begin{aligned}
&\langle \mathbf{N}, \eta | \hat{H}_{t-J} | \mathbf{N}, \eta \rangle = H_{t-J}(\mathbf{N}, \bar{\eta}, \eta) \\
&= \frac{Js^2}{2} \sum_{(i,j)} \mathbf{N}_i \cdot \mathbf{N}_j + \frac{J(s-\frac{1}{2})^2}{2} \sum_{(i,j)} \mathbf{N}_i \cdot \mathbf{N}_j \bar{\eta}_i \eta_i \bar{\eta}_j \eta_j \\
&+ \frac{Js(s-\frac{1}{2})}{2} \sum_{(i,j)} \mathbf{N}_i \cdot \mathbf{N}_j (\bar{\eta}_i \eta_i + \bar{\eta}_j \eta_j) \\
&- \frac{J(s+\frac{1}{2})^2}{2} \sum_{(i,j)} \bar{\eta}_i \eta_i \bar{\eta}_j \eta_j - 2JM(s+\tfrac{1}{2})^2 \\
&+ 2st \sum_{(i,j)} \Big(\cos\frac{\vartheta_i}{2} \cos\frac{\vartheta_j}{2} \\
&+ e^{i(\varphi_i - \varphi_j)} \sin\frac{\vartheta_i}{2} \sin\frac{\vartheta_j}{2} \Big) \bar{\eta}_i \eta_j \ .
\end{aligned} \tag{41}$$

The overlap of two states, with $\mathbf{N}' \approx \mathbf{N}$

$$\begin{aligned}
\langle \mathbf{N}', \eta' | \mathbf{N}, \eta \rangle &= {}_s\langle \mathbf{N}' | \mathbf{N} \rangle_s \Big[1 + \bar{\eta}' \eta \Big({}_{\frac{1}{2}}\langle \mathbf{N}' | \mathbf{N} \rangle_{\frac{1}{2}} \Big)^{-1} \Big] \\
&\simeq \exp\Big\{ is\Phi(\mathbf{N}', \mathbf{N}, \mathbf{N}_0) + e^{-\frac{i}{2}\Phi(\mathbf{N}', \mathbf{N}, \mathbf{N}_0)} \bar{\eta}' \eta \Big\} \ ,
\end{aligned} \tag{42}$$

where the phase $\Phi(\mathbf{N}', \mathbf{N}, \mathbf{N}_0)$ can be written as

$$\Phi(\mathbf{N}', \mathbf{N}, \mathbf{N}_0) = \mathbf{A}(\mathbf{N}) \cdot (\mathbf{N}' - \mathbf{N}) \tag{43}$$

and the vector potential $\mathbf{A}(\mathbf{N})$ is given as a solution to the equation

$$\nabla_{\mathbf{N}} \times \mathbf{A}(\mathbf{N}) = \mathbf{N} \ . \tag{44}$$

The continuum limit is given by

$$\begin{aligned}
Z = \int \mathcal{D}\mu_{s-\frac{1}{2}}(\mathbf{N}) \mathcal{D}\bar{\eta} \mathcal{D}\eta \exp\Big\{ \int_0^\beta d\tau \sum_i \big[i(s - \tfrac{1}{2}\bar{\eta}_i \eta_i) \mathbf{A}(\mathbf{N}_i) \cdot \dot{\mathbf{N}}_i \\
- \bar{\eta}_i \partial_\tau^{(\gamma)} \eta_i + \bar{\eta}_i \eta_i \big] - \int_0^\beta d\tau H_{t-J}(\mathbf{N}, \bar{\eta}, \eta) \Big\}
\end{aligned} \tag{45}$$

$$\int_0^\beta d\tau \bar{\eta}_i(\tau)\partial_\tau^{(\gamma)}\eta_i(\tau) \equiv \sum_{j=0}^{\mathcal{N}-1} \bar{\eta}_i(\tau_j)[\gamma_s \eta_i(\tau_j) - \eta_i(\tau_{j+1})] \ . \quad (46)$$

This expression for the partition function can be used as a starting point for the semiclassical approximation which we may apply if $s \to \infty$.

7.5 Discussion and Conclusions

We have shown in the present paper, how the graded Holstein-Primakoff representation can be constucted in the natural way for operators entering the Hamiltonian of the $t-J$ model, which is one of the most popular models used to describe the strongly correlated electron system. This representation allows to develop a systematic semiclassical approximation similar to spin-wave theory of the localized magnetism. Since the $t - J$ model describes the itinerant magnetism and has holes, this approximation is a semiclassical description of these holes interacting with the spin-waves. In the case of the square bipartite lattice one way to proceed is to divide it into two sublattices and act exactly as we do in the case of the Heisenberg antiferromagnet on the square lattice [10]. On one sublattice, say A, we will use the representation (26) on another sublattice, say B, we will use a unitary transformed representation

$$\begin{aligned} \hat{S}_0 &= s + \tfrac{1}{2}\hat{f}^\dagger \hat{f} \ , & \hat{X}_1 &= \hat{f}^\dagger \hat{b} \ , \\ \hat{S}^z &= -s + \hat{b}^\dagger \hat{b} + \tfrac{1}{2}\hat{f}^\dagger \hat{f} \ , & \hat{X}_2 &= -\hat{f}^\dagger \sqrt{2s - \hat{b}^\dagger \hat{b} - \hat{f}^\dagger \hat{f}} \ , \\ \hat{S}^+ &= \hat{b}^\dagger \sqrt{2s - \hat{b}^\dagger \hat{b} - \hat{f}^\dagger \hat{f}} \end{aligned} \quad (47)$$

and as a vacuum state the state $|\tilde{\Phi}_0\rangle = |s, -s, -s\rangle$. Then expanding the operators (26) and these operators in $1/s$ up to $O(1)$ and substituting them into $t - J$ model Hamiltonian (6) we will obtain the Hamiltonian describing a hole interacting with the spin waves in the Néel background. Such a Hamiltonian was proposed in [15] and the analysis of its spectrum was carried out.

A semiclassical description for other classical backgrounds can be obtained from the supercoherent states representation of the partition function for the $t - J$ model (45) as a path integral, which is another result of this paper.

Acknowledgments

It is a pleasure to thank Y. Gündüç for his invitation to visit the Department of Physics Engineering of Hacettepe University and for interesting discussions. A part of this work was done there. This work was done within the framework of the Associateship Scheme of the Abdus Salam International Centre for Theoretical Physics, Trieste, Italy. I would also like to express my gratitude to the Abdus Salam ICTP, where this work was finalized, for hospitality and providing a very stimulating atmosphere.

Support from the Scientific and Technical Research Council of Turkey (TÜBİTAK) within the framework of the NATO-CP Advanced Fellowship Programme is also gratefully acknowledged.

References

[1] P.W.Anderson, *Science* **235** (1987) 1196.

[2] J.Hubbard, *Proc. Phys. Soc.* **A276** (1963) 238.

[3] A.P.Balachandran, E.Ercolessi, G.Morandi and A.M.Srivastava, *The Hubbard Model and Anyon Superconductivity.* WS, 1990.

[4] S.Azakov, *Two Dimensional Hubbard Model and Heisenberg Antiferromagnet.* Lecture Notes IASBS, Zanjan, 1997.

[5] Note that the constraint (22) is needed only in order to get $K_2 = K_3 = 0$.

[6] F.Zhang and T.M.Rice, *Phys. Rev.* B**37** (1988) 3759.

[7] P.Wiegmann, *Phys. Rev. Lett.* **60** (1988) 821.

[8] T.Holstein and H.Primakoff, *Phys. Rev.* **58** (1940) 1098.

[9] J.D.Reger and A.P.Young, *Phys. Rev.* B**37** (1988) 5978.

[10] P.W.Anderson, *Phys. Rev.* **86** (1952) 694.
R.Kubo, *Phys. Rev.* **87** (1952) 568, *Rev. Mod. Phys.* **25** (1953) 344.
T.Oguchi, *Phys. Rev.* **117** (1960) 117.

[11] F.D.M. Haldane., *Phys. Lett.* A**93** (1983) 464; *Phys. Rev. Lett.* **50** (1983) 1153.

[12] M.Scheunert, W.Nahm and V.Rittenberg, *J. Math. Phys.***18** (1977) 155.

[13] J.M. Radcliffe, *J. Phys.* A**4** (1971) 313.

[14] E.Fradkin, *Field Theories of Condensed Matter Systems* . Addison-Wesley Publication, 1991.

[15] C.L.Kane, P.A.Lee and N.Read, *Phys. Rev.* B**39** (1989) 6880.

TUNING THE MAGNETISM OF ORDERED AND DISORDERED STRONGLY-CORRELATED ELECTRON NANOCLUSTERS

Nicholas Kioussis, Yan Luo, and Claudio Verdozzi

Department of Physics
California State University Northridge, California 91330-8268

Abstract Recently, there has been a resurgence of intense experimental and theoretical interest on the Kondo physics of nanoscopic and mesoscopic systems due to the possibility of making experiments in extremely small samples. We have carried out exact diagonalization calculations to study the effect of energy spacing Δ in the conduction band states, hybridization, number of electrons, and disorder on the ground-state and thermal properties of strongly-correlated electron nanoclusters. For the ordered systems, the calculations reveal for the first time that Δ tunes the interplay between the *local* Kondo and *non local* RKKY interactions, giving rise to a "Doniach phase diagram" for the nanocluster with regions of prevailing Kondo or RKKY correlations. The interplay of Δ and disorder gives rise to a Δ versus concentration $T = 0$ phase diagram very rich in structure. The parity of the total number of electrons alters the competition between the Kondo and RKKY correlations. The local Kondo temperatures, T_K, and RKKY interactions depend strongly on the local environment and are overall *enhanced* by disorder, in contrast to the hypothesis of "Kondo disorder" single-impurity models. This interplay may be relevant to experimental realizations of small rings or quantum dots with tunable magnetic properties.

Keywords: Phase diagram, Kondo effect, RKKY interaction, Nanoclusters.

8.1 Introduction

Magnetic impurities in non-magnetic hosts have been one of the central subjects in the physics of strongly correlated systems for the past four decades[1]. Such enduring, ongoing research effort is motivated by a constant shift and increase of scientific interest over the years, from dilute [2] to concentrated impurities [3], from periodic [4] to disordered samples [5, 6], and from macroscopic [7] to nanoscale phenomena

S. Halilov (ed.), Physics of Spin in Solids: Materials, Methods and Applications, 115–138.

[8]. Macroscopic strongly correlated electron systems at low temperatures and as a function of magnetic field, pressure, or alloying show a wide range of interesting phenomena, such as non-Fermi-liquid behavior, antiferromagnetism, ferromagnetism, enhanced paramagnetism, Kondo insulators, quantum criticality or superconductivity[1, 7]. These phenomena are believed to arise through the interplay of the Kondo effect, the electronic structure and intersite correlations. In the simplest single-impurity case, the Kondo problem describes the antiferromagnetic interaction, J, between the impurity spin and the free electron spins giving rise to an anomalous scattering at the Fermi energy, leading to a large impurity contribution to the resistivity[1]. The low-energy transport and the thermodynamic properties scale with a single characteristic energy, the Kondo temperature, $T_K \propto exp(-1/\rho(E_F)J)$, where $\rho(E_F)$ is the density of states of the conduction electrons at the Fermi energy [1]. At $T >> T_K$, the impurity spin is essentially free and the problem can be treated perturbatively. At $T << T_K$, the impurity spin is screened forming a singlet complex with the conduction electrons, giving rise to a local Fermi liquid state.

For dense Kondo or heavy fermion compounds containing a periodic array of magnetic ions interacting with the sea of conduction electrons, the physics is determined from the competition between the *non local* Ruderman-Kittel-Kasuya-Yosida (RKKY) interactions and the *local* Kondo interactions[9]. The RKKY interaction is an indirect magnetic interaction between localized moments mediated by the polarized conduction electrons, with an energy scale of order $J_{RKKY} \propto J^2\rho(E_F)$, which promotes long- or short-range magnetic ordering. On the other hand, the Kondo effect favors the formation of singlets resulting in a non-magnetic ground state. In the high temperature regime the localized moments and the conduction electrons retain their identities and interact weakly with each other. At low-temperatures, the moments order magnetically if the RKKY interaction is much larger than the Kondo energy, while in the reverse case, the system forms a heavy Fermi liquid of quasiparticles which are composites of local moment spins bound to the conduction electrons[7, 9]. Thus, the overall physics can be described by the well-known "Doniach phase diagram", originally derived for the simple Kondo necklace model[10]. The description of the low-temperature state, when both the RKKY and the Kondo interactions are of comparable magnitude, is an intriguing question that remains poorly understood and is the subject of active research[9].

The interplay of disorder and strong correlations has been a subject of intensive and sustained research, in view of the non-Fermi-liquid (NFL) behavior in the vicinity of a quantum critical point[11]. Various disorder-driven models have been proposed to explain the experimentally observed[7] NFL behavior at low temperatures[5–7, 12]. The phenomenological "Kondo disorder" approaches [5, 13], based on single-impurity models, assume a distribution of Kondo temperatures caused by a distribution of either $f - c$ orbital hybridization or of impurity energy levels. These models rely on the presence of certain sites with very low T_K spins leading to a NFL behavior at low T. An open issue in such single-site Kondo approaches, is whether the inclusion of RKKY interactions would renormalize and eliminate the low-T_K spins[4, 14–16]. An alternative view is the formation of large but finite magnetic clusters (Griffith phases) within the disordered phase through the competition between the RKKY and Kondo interactions [6, 17].

On the other hand, the possibility of making experiments in extremely small samples has lead to a resurgence of both experimental and theoretical interest of the physics of the interaction of magnetic impurities in nanoscopic and mesoscopic non-magnetic metallic systems. A few examples include quantum dots[18], quantum boxes[19] and quantum corrals[20]. Recent scanning tunneling microscope(STM) experiments [21] studied the interaction of magnetic impurities with the electrons of a single-walled nanotube confined in one dimension. Interestingly, in addition to the bulk Kondo resonance new sub peaks were found in shortened carbon nanotubes, separated by about the average energy spacing, Δ, in the nanotube. The relevance of small strongly correlated systems to quantum computation requires understanding how the infinite-size properties become modified at the nanoscale, due to the finite energy spacing Δ in the conduction band which depends on the size of the particle [8, 19, 22–24]. For such small systems, controlling T_K upon varying Δ is acquiring increasing importance since it allows to tune the cluster magnetic behavior and to encode quantum information. While the effect of Δ on the single-impurity Anderson or Kondo model has received considerable theoretical[8, 19, 22–24] and experimental[21] attention recently, its role on *dense* impurity clusters remains an unexplored area thus far. The low-temperature behavior of a nanosized heavy-electron system was recently studied within the mean-field approximation[25]. A central question is what is the effect of Δ on the interplay between the Kondo effect and the RKKY interaction

In this work we present exact diagonalization calculations for d- or f-electron nanoclusters to study the effects of energy spacing, parity of number of electrons, and hybridization on the interplay between Kondo and RKKY interactions in both *ordered* and *disordered* strongly correlated electron nanoclusters. While the properties of the system depend on their geometry and size[26], the present calculations treat exactly the

Kondo and RKKY interactions, the disorder averages, and they provide a distribution of local T_K's renormalized by the intersite f-f interactions. Our results show that: i) tuning Δ and the parity of the total number of electrons can drive the nanocluster from the Kondo to the RKKY regime, i.e. a zero- temperature energy spacing versus hybridization phase diagram; ii) the temperature versus hybridization "Doniach" phase diagram for nanoclusters depends on the energy spacing ; iii) changing the total number of electrons from even to odd results in an enhancement (suppression) of the local Kondo (RKKY) spin correlation functions; iv) the Δ versus alloy concentration $T = 0$ phase diagram exhibits regions with prevailing Kondo or RKKY correlations alternating with domains of ferromagnetic (FM) order; and v) the local T_K's and the nearest-neighbor (n.n) RKKY interactions depend strongly on the local environment and are overall *enhanced* by disorder. The disorder-induced enhancement of T_K in the clusters is in contrast to the hypothesis of "Kondo disorder" models for extended systems.

The rest of the paper is organized as follows. In Sec. II, we describe the model for both the periodic and disordered clusters. In Secs. IIIA and IIIB we present results for the ground-state and thermal properties of the ordered and disordered nanoclusters, respectively. Section IV contains concluding remarks.

8.2 Methodology

The one dimensional Anderson lattice model is

$$H = -t\sum_{i\sigma}(c^{\dagger}_{i\sigma}c_{i+1\sigma} + H.c) + E_f\sum_{i\sigma} n^f_{i\sigma} \\ +U\sum_{i} n^f_{i\uparrow}n^f_{i\downarrow} + V\sum_{i\sigma}(f^{\dagger}_{i\sigma}c_{i\sigma} + H.c.). \quad (1)$$

Here, t is the nearest-neighbor hopping matrix element for the conduction electrons, $c^{+}_{i,\sigma}(c_{i,\sigma})$ and $f^{+}_{i,\sigma}(f_{i,\sigma})$ create (annihilate) Wannier electrons in c- and f- like orbital on site i with spin σ, respectively; E_f is the energy level of the bare localized orbital, V is the on-site hybridization matrix element between the local f orbital and the conduction band and U is the on-site Coulomb repulsion of the f electrons. We use a simple nearest-neighbor tight-binding model for the conduction band dispersion, $\epsilon_k = -2tcosk$. We consider the half-filled ($N_{el} = 2N$) symmetric ($E_f = -\frac{U}{2}$) case, with $U = 5(6)$ for the periodic (disorder) case. We investigate one-dimensional rings of $N = 4$ and 6. Most of the results presented are for the $N = 6$ case, except for the results for $T > 0$ where we have used $N = 4$ sites. The exact diagonalization calculations employ periodic boundary conditions.

Ordered Clusters

We have investigated the ground-state properties as a function of the hybridization and the energy spacing in the conduction band, $\Delta = 4t/(N-1) = \frac{4t}{5}$. We have calculated the average $f-$ and $c-$local moments, $< (\mu_i^f)^2 > \equiv < S_i^{f,z} S_i^{f,z} >$ and $< (\mu_i^c)^2 > \equiv < S_i^{c,z} S_i^{c,z} >$, respectively. Here, S_i^f and S_i^c are given by

$$S_i^f = \frac{1}{2} \sum_{\sigma,\sigma'} \tau_{\sigma\sigma'} f_{i\sigma}^+ f_{i\sigma'} \tag{2}$$

and

$$S_i^c = \frac{1}{2} \sum_{\sigma,\sigma'} \tau_{\sigma\sigma'} c_{i\sigma}^+ c_{i\sigma'}, \tag{3}$$

where τ are the Pauli matrices.

We have also calculated the zero-temperature *f*-*f* and *f*-*c* spin correlation functions (SCF) $< S_i^f S_{i+1}^f > \equiv < g | S_i^{f,z} S_{i+1}^{f,z} | g >$ and $< S_i^f S_i^c > \equiv < g | S_i^{f,z} S_i^{c,z} | g >$, respectively. Here, $|g>$ is the many-body ground state and $S_i^{f,z}$ is the z-component of the f-spin at site i. As expected, the cluster has a singlet ground state ($S_g = 0$ where S_g is the ground-state spin) at half filling. We compare the onsite Kondo correlation function $< S_i^f S_i^c >$ and the nearest-neighbor RKKY correlation function $< S_i^f S_{i+1}^f >$ to assign a state to the Kondo or RKKY regimes, in analogy with mean field treatments[27]. The spin structure factor related to the equal-time $f-f$ spin correlation functions in the ground state is

$$S^{ff}(q) = \frac{1}{N} \sum_{i,j} < g | \mathbf{S}_i^f \cdot \mathbf{S}_j^f | g > e^{iq(x_i - x_j)}. \tag{4}$$

The temperature-dependent local f-spin susceptibility, $\chi^f(T)$,is

$$\frac{k_B T \chi^f(T)}{(g\mu_B)^2} = \frac{1}{Q} \sum_{\alpha} e^{-\frac{E_\alpha}{k_B T}} < \alpha | S^f(i) S^{Tot} | \alpha >, \tag{5}$$

where

$$Q = \sum_{\alpha} e^{-\frac{E_\alpha}{k_B T}} \tag{6}$$

is the partition function. Here, S^{Tot} is the z-projection of the total spin (both the *f*- and *c*-contributions), and $|\alpha>$ and E_α are the exact many-body eigenstates and eigenvalues, respectively. When $V = 0$, the localized spins and conduction electrons are decoupled and $\chi^f(T)$ is simply the sum of the Curie term due to the free f spins and the Pauli term of the free conduction electrons. For finite V, $\chi^f(T)$ decreases with temperature at low-temperatures. The specific heat is calculated from the second derivative of the free energy F, $C_v = -T\frac{\partial^2 F}{\partial T^2}$. At $V = 0$, the specific heat of the system is given by the sum of the delta function at $T = 0$ that originates from the free localized spins and the specific heat of free conduction electrons. For finite V the specific heat exhibits a double-peak structure: the high-temperature peak is almost independent of the hybridization and arises from the free conduction electron contribution, whereas the low-temperature peak varies strongly with hybridization.

Disordered Clusters

We consider a random binary alloy cluster, $A_{N-x}B_x$, of N=6 sites and different number of B atoms, x = 0-N, arranged in a ring described by the half-filled ($N_{el} = 2N = 12$) two-band lattice Anderson Hamiltonian in Eq.(1). We introduce binary disorder in the f-orbital energy ϵ_f^i ($= \epsilon_f^A$ or ϵ_f^B) and in the intra-atomic Coulomb energy U_i ($= U^A$ or U^B), to model a Kondo-type A atom with $\epsilon_f^A = -U^A/2 =$ -3 (symmetric case) and a mixed-valent (MV) type B atom with ϵ_f^B = -2 and $U^B = 1$. For both types of atoms $V^A = V^B = V = 0.25$. For $t = 1$, this choice of parameters leads to a degeneracy between the doubly-degenerate c-energy levels, $\epsilon_k = -t$, and the energy level $\epsilon_f^B + U^B$. Upon filling the single particle energy levels for any x, $N - x$ (x) electrons fill the ϵ_f^A (ϵ_f^B) levels, and two electrons the -2t conduction energy level, with the remaining $N - 2$ electrons accommodated in the x+4 degenerate states at $-t$. This in turn results in strong charge fluctuations.

The temperature-dependent f susceptibility, $\chi_x^f(T)$, at concentration x, is given by

$$\frac{k_B T \chi_x^f(T)}{(g\mu_B)^2} = \frac{1}{Q} \sum_{C_x, \alpha_{C_x}} e^{-\frac{E_{\alpha C_x}}{k_B T}} < \alpha_{C_x} | S^f(i) S^{Tot} | \alpha_{C_x} >, \qquad (7)$$

where C_x denote the configurations at concentration x, $|\alpha_{C_x} >$ and E_{C_x} are the configuration-dependent exact many-body eigenstates and eigenvalues, respectively, and Q denotes the partition function.

8.3 Results and discussion

Ordered Clusters

1. Ground State Properties

In Fig. 1 we present the variation of the local Kondo SCF $< S_i^f S_i^c >$ (squares) and the nearest-neighbor RKKY SCF $< S_i^f S_{i+1}^f >$ (circles) as a function of hybridization for two values of the hopping matrix element $t = 0.2$ (closed symbols) and $t = 1.2$ (open symbols), respectively. As expected, for weak hybridization V the local nearest-neighbor RKKY (Kondo) SCF is large (small), indicating strong short-range antiferromagnetic coupling between the $f - f$ local moments, which leads to long range magnetic ordering for extended systems. As V increases, $< S_i^f S_{i+1}^f >$ decreases whereas the $< S_i^f S_i^c >$ increases (in absolute value) saturating at large values of V. This gives rise to the condensation of independent local Kondo singlets at low temperatures, i.e., a disordered spin liquid phase. For large V the physics are *local*. Interestingly, as t or Δ decreases the f-c spin correlation function is dramatically enhanced while the f-f correlation function becomes weaker, indicating a transition from the RKKY to the Kondo regime.

In Fig. 2 we present the average local f- (circles) and c- (squares) moments as a function of hybridization for two values of the hopping matrix element $t = 0.2$ (closed symbols) and $t = 1.2$ (open symbols), respectively. In the weak hybridization limit, the large on-site Coulomb repulsion reduces the double occupancy of the f level and a well-defined local f moment is formed $\langle \mu_f^2 \rangle = 1.0$ while $\langle \mu_c^2 \rangle = 0.5$. With increasing V both charge- and spin- fluctuations become enhanced and the local $f-$ moment decreases monotonically whereas the $c-$ local moment exhibits a maximum. In the large V limit both the $f-$ and $c-$ local moments show similar dependence on V, with $< \mu_c^2 > \approx < \mu_f^2 >$, indicating that the *total* local moment μ vanishes. The effect of lowering the energy spacing Δ is to decrease (increase) the $f-$ ($c-$) local moment, thus tuning the magnetic behavior of the system. Note that the maximum value of the $c-$ local moment increases as Δ decreases. This is due to the fact that for smaller t values the kinetic energy of conduction electrons is lowered, allowing conduction electrons to be captured by f electrons to screen the local f moment, thus leading to an enhancement of the local $c-$ moment.

In Fig. 3 we present the energy spacing versus V zero-temperature phase diagram of the nanocluster, which illustrates the interplay between Kondo and RKKY interactions. In the RKKY region $< S_i^f S_{i+1}^f >$ is larger than the $< S_i^f S_i^c >$ and the *total* local moment is non zero; in

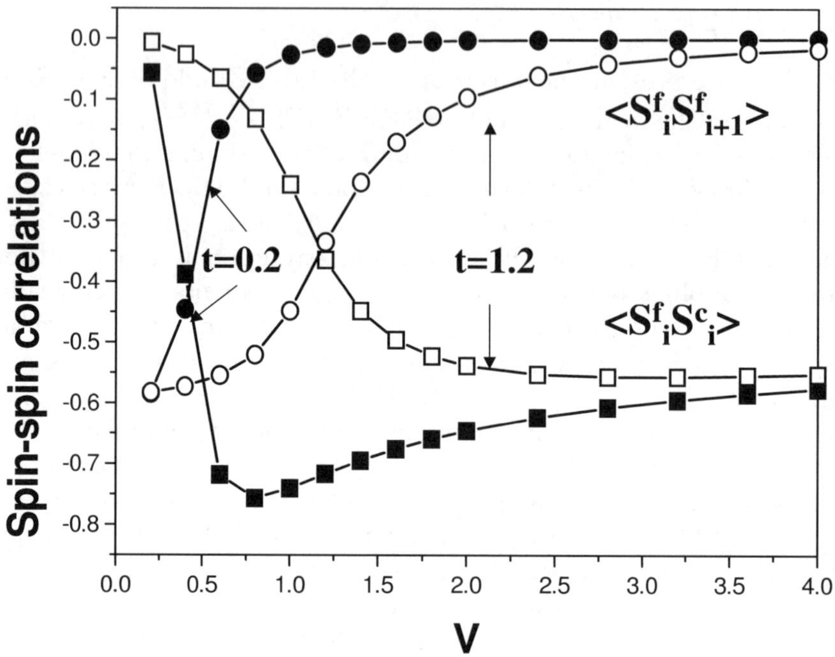

Figure 1. Nearest neighbor f-f spin-spin correlations (circles) and on-site f-c spin-spin correlations (squares) as a function of V for two values of the hopping parameter of $t = 0.2$ (closed symbols) and $t = 1.2$ (open symbols), respectively.

the Kondo regime $< S_i^f S_{i+1}^f >$ is smaller than the $< S_i^f S_i^c >$, the *total* local moment vanishes, and the ground state of the system is composed of independent local singlets. The solid crossover curve indicates the $V = V_c$ or $\Delta = \Delta_c$ values, where the *local* and *non local* spin correlation functions are equal, i.e., $< S_i^f S_{i+1}^f >=< S_i^f S_i^c >$. The dashed curve denotes the set of points where the on-site *total* local moment $\mu = 0$. Thus, in the intermediate regime, which will be referred to as the *free spins* regime [11], $< S_i^f S_{i+1}^f >$ is smaller than the $< S_i^f S_i^c >$, the f moment is *partially* quenched and $\mu \neq 0$. Interestingly, we find that the *free spins* regime becomes narrower as the average level spacing Δ is reduced. This result may be interpreted as a quantum critical regime (QCP) for the nanoring due to the finite energy spacing, which eventually reduces to a quantum critical point when $\Delta \to 0$.

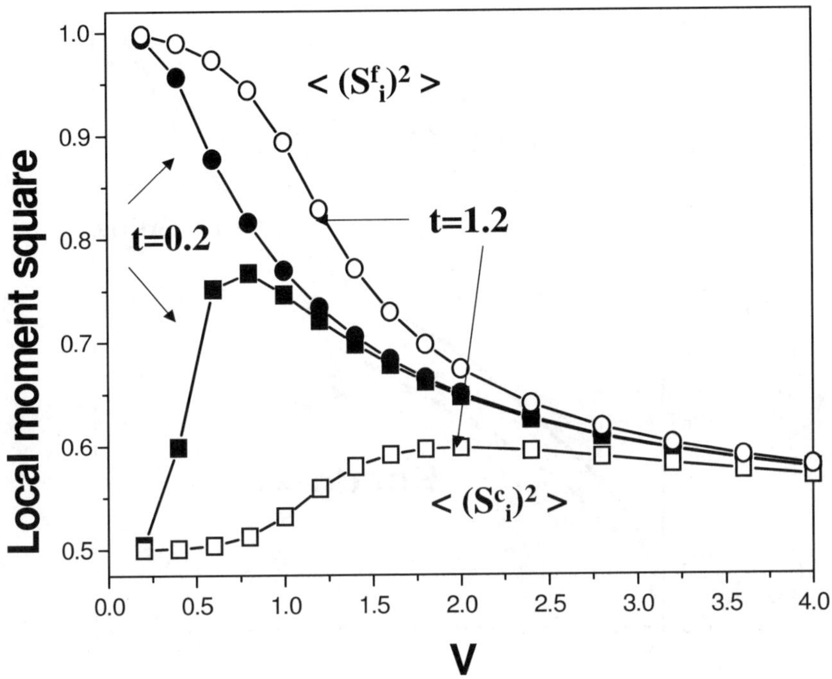

Figure 2. f- (circles) and $c-$ (squares) local moment versus hybridization for two values of the hopping parameter of t=0.2 (closed symbols) and t=1.2 (open symbols), respectively.

Fig. 4 shows the spin structure factor of the local f electrons $S^{ff}(q)$ for various values of V and for $t = 0.2$. As discussed earlier, the ground state of the half-filled symmetric periodic Anderson model is a singlet. For small V, the spin structure factor exhibits a maximum at $q = \pi$, indicating the presence of strong antiferromagnetic correlations between the local f moments, consistent with the large values of $< S_i^f S_{i+1}^f >$ in Fig. 1. With increasing hybridization, the maximum of $S^{ff}(q = \pi)$ decreases and vanishes at very large hybridization, indicating that the ground state undergoes a transition from the antiferromagnetic to the nonmagnetic Fermi liquid phase. This is consistent with the zero-temperature phase diagram in Fig. 3.

The spin gap as a function of hybridization V for two values of energy spacing is shown in Fig. 5. The spin gap is defined as the energy

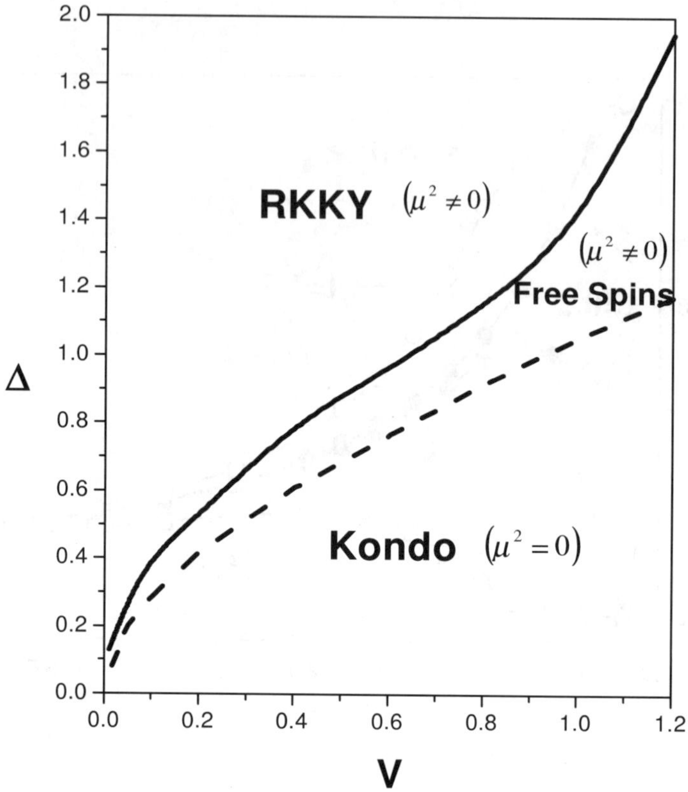

Figure 3. Energy spacing Δ versus hybridization zero-temperature phase diagram. The solid curve denotes the crossover point of the spin-spin correlation function in Fig. 1; the dashed curve denotes the set of points where the on-site *total* moment square $\langle(\mu_f + \mu_c)^2\rangle = 0.0 \pm 0.05$.

difference between the singlet ground state and the lowest-lying excited triplet ($S = 1$) state. As expected, there is a nonzero spin gap for the half-filled Anderson lattice model, which increases with hybridization. Interestingly, the spin gap dramatically increases as the average energy level spacing Δ is reduced. Thus, the energy spacing or equivalently the size of the cluster tunes the low-energy excitation energy which controls the low-temperature specific heat and susceptibility.

2. Thermal Properties

The T=0 exact diagonalization results on small clusters are generally plagued by strong finite size effects[26, 28]. Performing calculations at

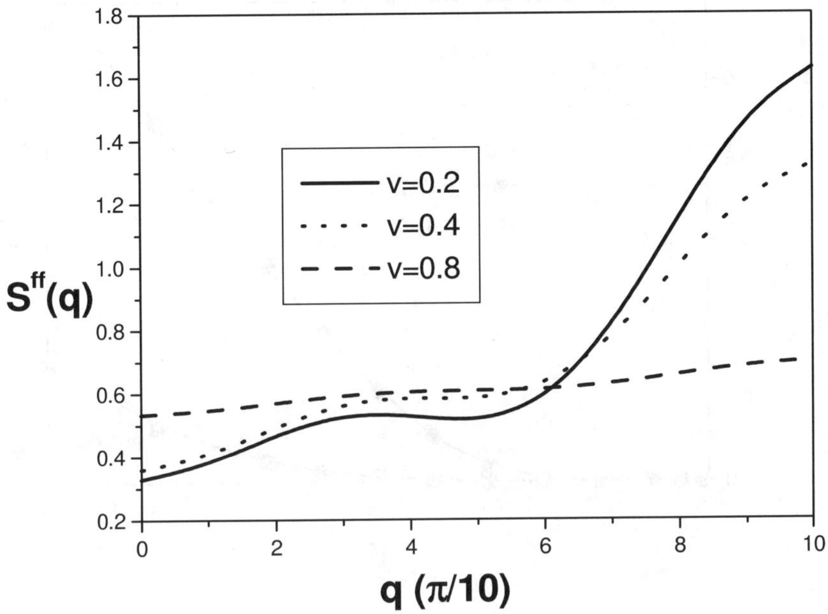

Figure 4. Spin structure factor as a function of wave-vector for different values of V and for t=0.2.

$T > 0$ gives not only the thermodynamic properties of the system, but most importantly diminishes finite-size effects for ($k_B T \gg \Delta$).

In Fig. 6, we show the nearest-neighbor f-f spin-spin correlations and on-site f-c spin-spin correlation as a function of temperature for for $t = 0.2$ and for $V = 0.2 < V_c$ and $V = 0.4 > V_c$, where $V_c = 0.25$. At high temperatures, the free moments of the f and conduction electrons are essentially decoupled. The nearest-neighbor *non local* spin correlation function falls more rapidly with T than the on-site *local* $f - c$ spin-spin correlations, indicating that the *non local* spin correlations can be destroyed easier by thermal fluctuations. For $V < V_c$, the nanocluster is dominated by RKKY (Kondo) interactions at temperatures lower (higher) than the crossover temperature, T^{cl}_{RKKY}, which denotes the temperature where the *non local* and *local* interaction become equal in the nanocluster. In the infinite system this temperature would denote the ordering Néel temperature. On the other hand, for $V > V_c$ the

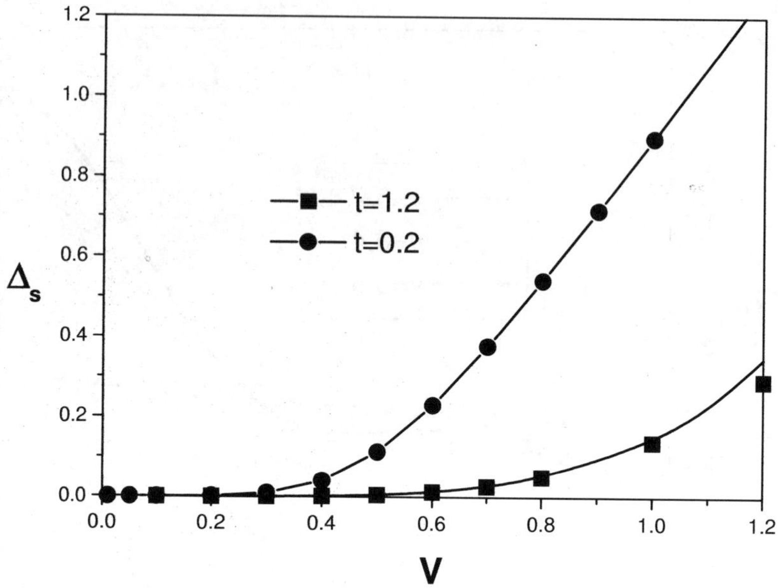

Figure 5. Spin gap as a function of V for $t = 0.2$ and 1.2. The spin gap increases exponentially (linearly) for small (large) V.

RKKY and Kondo spin correlation functions do not intersect at any T, and the physics become dominated by the local interactions.

In Fig. 7 we present the crossover temperature T^{cl}_{RKKY} for the cluster as a function of hybridization for different values of t. This represents the phase diagram of the strongly correlated nanocluster, which is similar to the "Doniach phase diagram" for the infinite Kondo necklace model. The phase within the crossover curve denotes the regime where the *non local* short-range magnetic correlations are dominant. For $V < V_c$ and $T >> T^{cl}_{RKKY}$ one enters into the disordered "free" local moment regime. On the other hand, for $V > V_c$ and at low T, the nanocluster can be viewed as a condensate of singlets, typical of the Kondo spin-liquid regime. Interestingly, the T^{cl}_{RKKY} can be tuned by the energy spacing Δ or the size of the cluster. Thus, increasing Δ or decreasing the size of the nanocluster results to enhancement of the *non local* nearest-neighbor

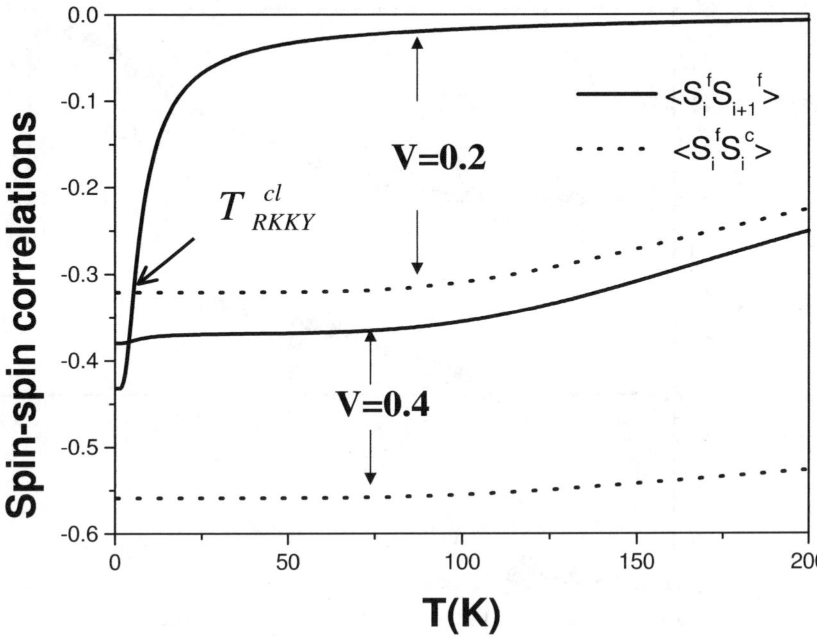

Figure 6. Nearest-neighbor f-f and on-site f-c spin-spin correlation functions versus temperature for $t = 0.2$ and for $V = 0.2 < V_c$ and $V = 0.4 > V_c$, $V_c = 0.25$.

magnetic correlations and hence T^{cl}_{RKKY}. This result is the first exact "Doniach phase diagram" for a nanocluster.

In bulk Kondo insulators and heavy-fermion systems, the low-T susceptibility and specific heat behavior is determined by the spin gap, which for the half-filled Anderson lattice model, is determined by the ratio of V to U. On the other hand, strongly correlated nanoclusters are inherently associated with a new low-energy cutoff, namely the energy spacing Δ of the conduction electrons. Thus, a key question is how can the low-temperature physics be tuned by the interplay of the spin gap and the energy spacing. In Fig. 8 we present the local f magnetic susceptibility as a function of temperature for $t = 0.2$ and for $V = 0.2 < V_c$, $V = V_c = 0.25$, and $V = 0.4 > V_c$. For small V, the spin gap which is smaller than Δ controls the exponential activation behavior of χ^f at low T. On the other hand, in the large V limit, the spin gap becomes larger than Δ (see Fig. 5) and the low-T behavior of the susceptibility

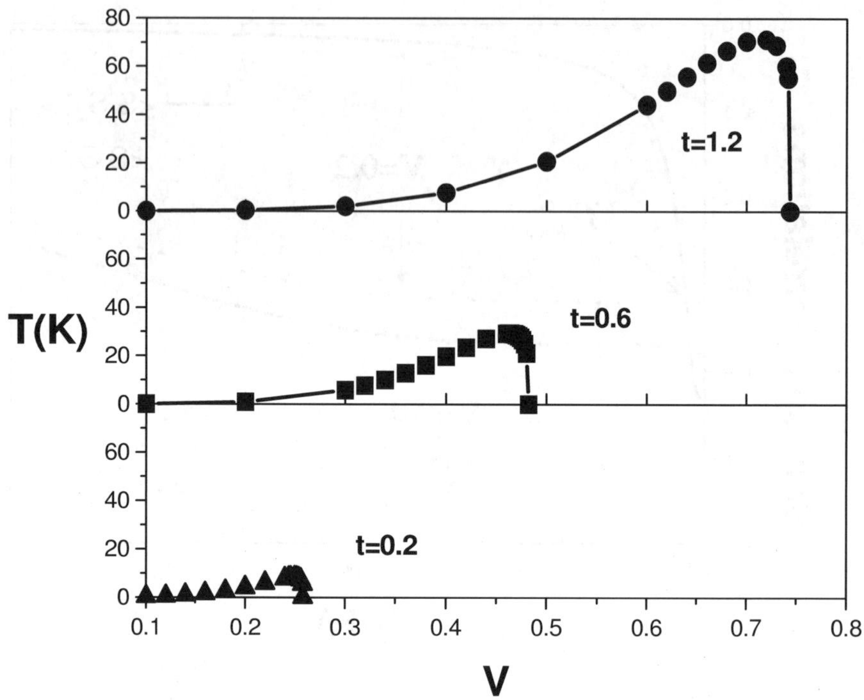

Figure 7. Effect of energy spacing, $\Delta = \frac{4t}{N-1}$ on the exact "Doniach phase diagram" for a strongly correlated electron nanochain. The crossover curve represents the crossover temperature T^{cl}_{RKKY}, where the non local short range AF spin correlations become equal to the local on-site Kondo spin correlations.

shows no exponential activation. At high T we can see an asymptotic Curie-Weiss regime, typical of localized decoupled moments.

In Fig. 9, we present the specific heat as a function of temperature for $V = 0.4$ and different t. At $V = 0$, the specific heat is given by the sum of a delta function at $T = 0$ for the localized spins and the specific heat of free fermions. As expected, by switching on the coupling V, they are combined to form a two-peak structure. The broad peak at high T is rather similar to the free-electron gas. The low-T behavior is associated with the lowest energy scale, which as in the case of the susceptibility, is determined by the lowest value between the spin gap and the energy spacing Δ. For large values of t (or Δ) the spin gap is reduced (see Fig. 5) and the spin gap is the lowest energy scale. Consequently, the low-T behavior exhibits exponential activation associated with the spin

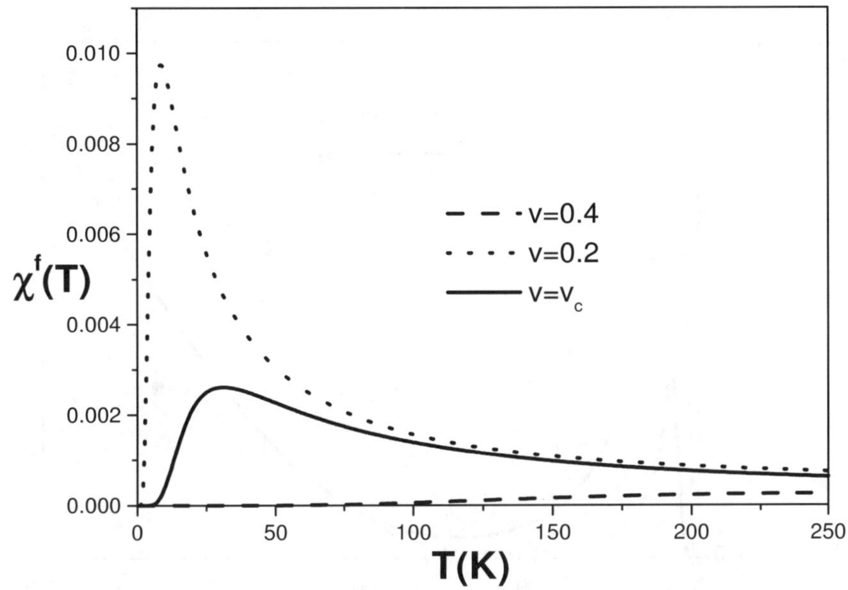

Figure 8. Local f magnetic susceptibility as a function of temperature for $t = 0.2$ and with $V = 0.2 < V_c$, $V = V_c = 0.25$ and $V = 0.4 > V_c$.

gap. On the other hand, for small energy spacing the physics become local (Kondo regime) and the low-T sharp peak shifts towards higher temperatures and becomes broader.

Disordered Clusters

1. Effect of Disorder

The configurations for $x \leq 3$ are shown in Fig. 10, left panel, along with the value of the spin, S_g, of the ground-state. The A (B) atoms are denoted by closed (open) circles, respectively. Except for the homogenous cases (x=0 and $x = 6$), with a $S_g = 0$ ground state, for all x there are configurations with $S_g \neq 0$. The average occupation and average LM for the periodic Kondo and MV lattices are $< n_f^A >= 1$, $< (\mu_f^A)^2 > = 0.99$, and $< n_f^B > = 1.6$, $< (\mu_f^B)^2 > = 0.43$, respectively. We carry out a detailed analysis for x=1 (S_g =2) to demonstrate the FM transition

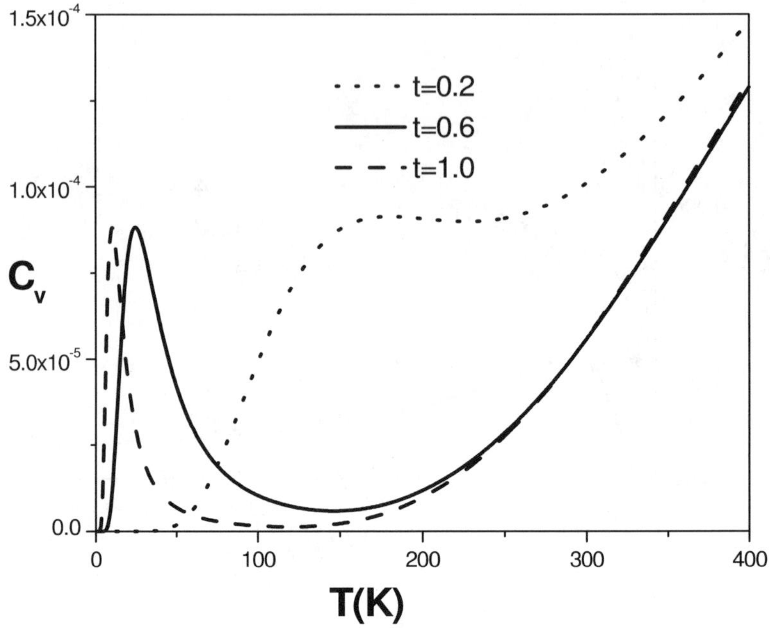

Figure 9. Specific heat as a function of temperature with $V = 0.4$ and various values of $t = 0.2$, 0.6 and 1.0. The low-T peak for larger energy spacing is due to the spin gap.

induced by a single MV atom in an otherwise Kondo cluster. Studies of extended systems have reported similar occurrence of ferromagnetism in the MV phase[29]. As expected, the singlet ground state of the $x = 0$ Kondo cluster is characterized by n.n. anti-ferromagnetic (AF) f-f spin correlations ($< S_f^A(i) S_f^A(i+1) >$ = - 0.58). The introduction of a MV atom renders them ferromagnetic. Since U_B is small, the B impurity tends to remove charge from the the conduction band, in particular from the k-state with $\epsilon_k = -t$, which has large amplitude at the B site and at the opposite A site across the ring. Such a depletion is different for the two spin states, thus yielding a maximum value for the f-moment of the MV atom. The f-f spin correlation function between the Kondo and MV atoms are AF ($< S_f^A(i) S_f^B(i+1) >$ = - 0.23), while they are FM among the Kondo atoms ($< S_f^A(i) S_f^A(i+1) > = +0.94$). A similar result was recently found in *ab initio* calculations[30], where introducing

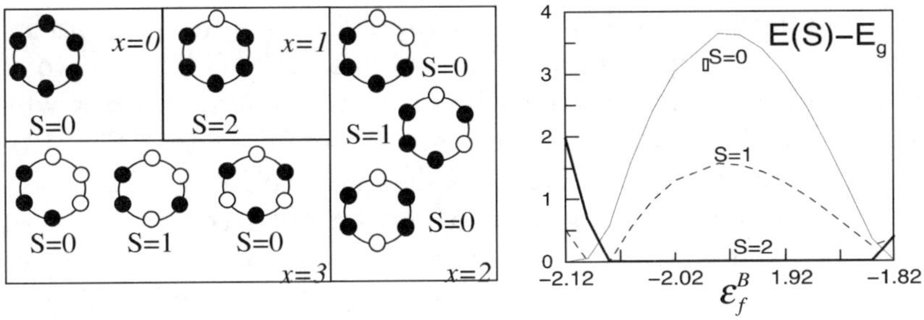

Figure 10. Left panel: Alloy configurations for various concentrations $x \leq 3$ (the $x > 3$ cases are obtained by exchanging closed and open circles). For each $x \leq 3$ configuration, the value of the ground-state spin S_g is reported. Right panel: Energy difference (in units of $10^{-4}t$) between the lowest $S \leq 2$ eigenstates and the ground state as function of ϵ_B.

a nitrogen impurity in small (1-5 atoms) Mn clusters induces ferromagnetism via AF coupling between the N to the Mn atoms, whilst Mn-Mn couple ferromagnetically. We find that there is a crossover in S_g from 0 $\rightarrow 1 \rightarrow 2 \rightarrow 0$ (Fig. 10, right panel) indicating a reentrant nonmagnetic transition around $\epsilon_B = 2$. This almost saturated FM $S_g = 2$ domain is robust against small changes in U_B, V, ϵ_A, U_A, cluster size ($N = 7$), and band filling ($N_{el} = 10$) provided that the Kondo atom has a large LM.

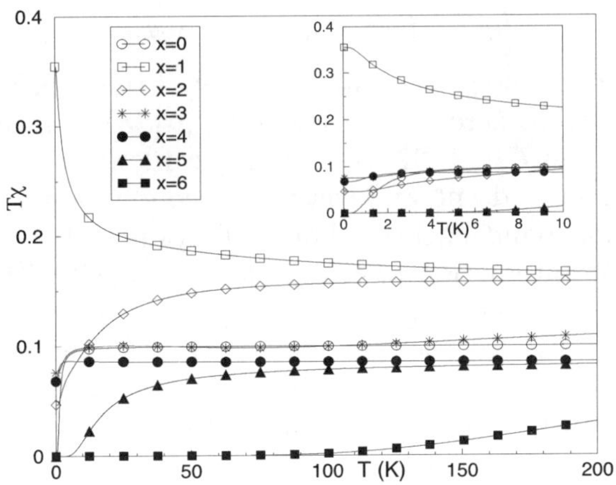

Figure 11. Temperature dependence of the average f-susceptibility for different alloy concentrations. The inset shows the low-temperature behavior.

In Fig. 11 we present $T\chi_x^f(T)$ as a function of temperature for different x. As $T \to 0$ (inset Fig. 11) $T\chi_x^f(T)$ approaches a finite value for $x = 1$–4 while it vanishes exponentially for x=0, 5 and 6. This is due to the fact that the former concentrations involve some configurations which are magnetic, while the latter have singlet ground states (Fig. 10). The stronger (weaker) low-temperature dependence for $x = 1$ ($x = 2 - 4$) is due to the smaller (larger) spin gap between the ground state and the lowest excited states. The magnetic susceptibility displays also a magnetic crossover upon varying x, and reveals a Curie-like divergence at low T for $x = 1 - 4$. The temperature-dependent results for the specific heat, not reported here, show corroborative evidence of this disorder-induced magnetic crossover.

2. Effect of Energy Spacing

Next we address a number of important open issues, namely (1) the effect of Δ on the interplay between RKKY and Kondo interactions in disordered clusters, (2) the characterization of the single-impurity "Kondo correlation energy" T_K in a *dense-impurity* cluster and (3) the effect of disorder and Δ on the distribution of the local T_K's. In the following, $\epsilon_B = -2$.

In contrast with previous studies, which introduced a phenomenological distribution $P(T_K)$ of single-impurity Kondo temperatures, the advantage of the present calculations is that one calculates exactly the Kondo correlation energy: we employ the so-called "hybridization" approach[31], with T_K defined as

$$k_B T_K(i) = E_g(V_i = 0) - E_g, \tag{8}$$

where $E_g(V_i = 0)$ is the ground-state energy of the dense-impurity cluster when V is set to zero at the i*th* site. Eq.(8) reduces to $k_B T_K = E_{band} - E_F + \epsilon_f - E_g$[22, 32] in the single impurity case. Here, E_F is the highest occupied energy level in the conduction band and E_{band} is the conduction band energy. This definition of the local T_K takes into account the interaction of the f-moment at site i with the other f-moments in the system [33].

In Table I we list for the periodic, x=0, case the local Kondo f-c spin correlation function $< S_f^A(i) S_c^A(i) >$, the n.n. f-f spin correlation function $< S_f^A(i) S_f^A(i+1) >$, and the local Kondo temperature for two different values of t (The energy spacing is $\Delta = 4t/(N-1) \equiv 4t/5$). As t or Δ decreases the f-c spin correlation function is dramatically enhanced while the f-f correlation function becomes weaker, indicating a transition from the RKKY to the Kondo regime. This is also corroborated by the increase in the local $T_K(i)$. The energy spacing affects not only the magnetic (A) atoms but the MV atoms as well. Thus, increasing t drives the B atoms from the non-magnetic, NM ($n_f \approx 2$), to the MV and finally to the Kondo regime.

We next examine the role of even versus odd number of electrons on the magnetic behavior of the uniform x=0 case. For $t = 1$, changing the number of electrons from $N_{el} = 12$ to $N_{el} = 11$ results in: (a) an enhancement of the local Kondo f-c spin correlation function, $< S_f^A(i) S_c^A(i) >$ from -0.01 to -0.12; and (b) a suppression of the nearest-neighbor f-f spin correlation function $< S_f^A(i) S_f^A(i+1) >$ from -0.58 to -0.20 (Due to the broken symmetry for $N_{el} = 11$, the f-f spin correlation functions range from -0.5 to +0.02). This interesting novel tuning of the magnetic behavior can be understood as follows: For an odd-electron cluster, the topmost occupied single particle energy level is singly occupied. On the other hand, for an even-electron cluster, the topmost occupied single-particle energy level is doubly occupied, thus blocking energy-lowering spin-flip transitions. This energy penalty intrinsically weakens the Kondo correlations[19]. As expected, changing the number of electrons from even to odd changes $S_g = 0$ to $S_g = \frac{1}{2}$. For t= 0.05 (Kondo regime), the on site f-c correlation function does not depend as strongly on the parity in the number of electrons because the sites are locked into singlets. On the other hand, $< S_f^A(i) S_f^A(i+1) >$ becomes suppressed as in the case of large energy spacing. Similar results were found for the various disordered concentrations.

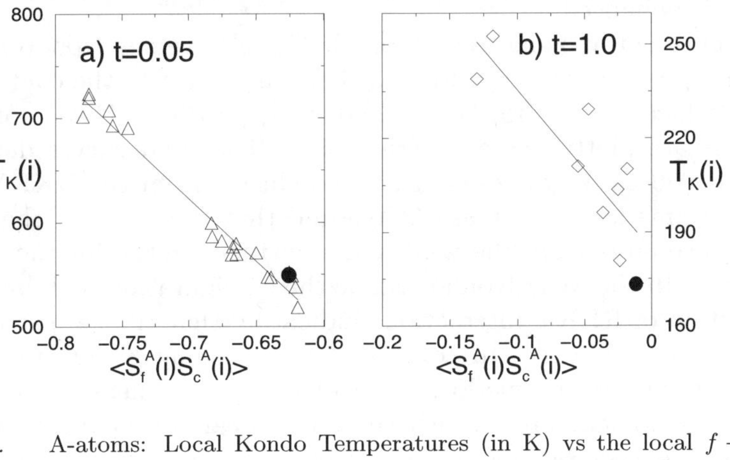

Figure 12. A-atoms: Local Kondo Temperatures (in K) vs the local $f - c$ spin correlation function, for different configurations and two different values of t. The closed circles refer to the $x = 0$ case and the lines are a guide to the eye.

In Fig. 12 we present the local $T_K(i)$ as a function of the local *f*-*c* spin correlation function $< S_f^A(i) S_c^A(i) >$ for all Kondo (A) atoms in the

Table 1. Local Kondo *f*-*c* and n.n. *f*-*f* spin correlations functions and the local Kondo temperature (in K) for two values of t (in eV). The average energy spacing is $\Delta = 4t/(N-1) \equiv 4t/5$.

	$< S_f^A(i) S_c^A(i) >$	$< S_f^A(i) S_f^A(i+1) >$	$T_K(i)$
t=0.05	-0.626	-0.322	551.8
t=1.00	-0.011	-0.584	173.4

singlet ground state at any concentration x for $t=$ 0.05 and 1.0. Note the different scales both on the horizontal and vertical axis in the panels. In both panels, the closed circles correspond to the x=0 lattice case and the line is a guide to the eye. The results indicate a correlation between T_K and the *f*-*c* spin correlation function (the larger T_K's correspond to the more negative *f*-*c* values) as one would expect, since both provide a measure of the Kondo effect. For t=0.05, most of the disordered cluster configurations are in the Kondo regime ($S_g = 0$), with larger T_K values; consequently, panel (a) has a larger number of singlet configurations. The introduction of MV impurities induces a distribution of $T_K(i)$'s, whose values are overall *enhanced* compared to those for the x=0 case, except for several configurations for t=0.05, in contrast with single-site theories for extended systems[5]. It is interesting that $P(T_K)$ for t=0.05 exhibits a bimodal behavior centered about 710 and 570K, respectively: The higher T_K's originate from isolated Kondo atoms which have MV atoms as n.n. so that the local screening of the magnetic moment of the A atom is enhanced.

The effect of alloying and Δ on the RKKY versus Kondo competition for a given x is seen in Fig. 13 (left panel), where the configuration averaged local $< S_f^A(i) S_c^A(i) >_x$ and $< S_f^A(i) S_f^A(i+1) >_x$ correlation functions are plotted as a function of t. The solid curves denote the uniform x=0 case, where we drive the cluster from the RKKY to the Kondo regime as we decrease t. We find that the stronger the average Kondo correlations are the weaker the average RKKY interactions and vice versa. In the weak Kondo regime the configurations exhibit a wider distribution of RKKY interactions indicating that they are sensitive to the local environment. In contrast, in the strong Kondo regime, the Kondo (A) atoms become locked into local Kondo singlets and the n.n. RKKY interactions are insensitive to the local environment. Interestingly, both energy spacing and disorder lead to an overall enhancement of the RKKY interactions compared to the homogenous state.

In the right panel of Fig. 13 we present the t versus x phase diagram for the nanocluster at $T = 0$. We compare the $< S_f^A(i) S_c^A(i) >_x$ and $< S_f^A(i) S_f^A(i+1) >_x$ to assign a state of specific concentration to

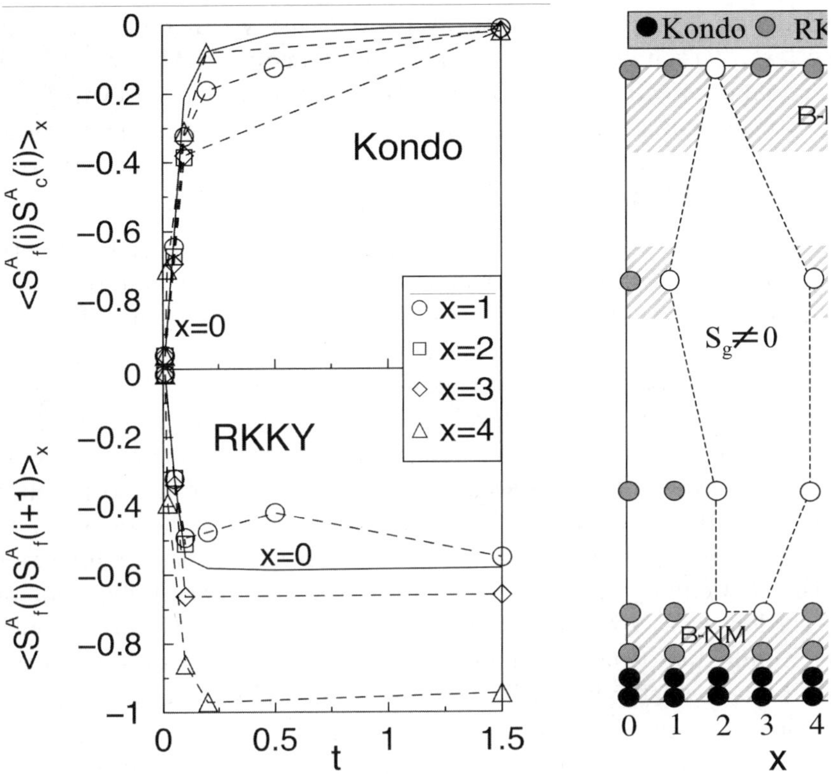

Figure 13. Left panel: Configuration-averaged local $f-c$ (top) and n.n. $f-f$ spin correlations (bottom) for the A atoms as function of t. The solid line refers to the homogenous $x=0$ case. Right panel: Zero-temperature t vs x phase diagram for the nanocluster. Black (gray) circles denote the Kondo (RKKY) regime. The white circles and the dashed contour delimit the FM region. The horizontal stripes denote the non-magnetic (NM), mixed valence (MV) and local moment (LM) behavior of the B-atoms.

the Kondo or RKKY regimes (black and gray circles, respectively), in analogy with the $x=0$ case (Table I) and with mean field treatments [27]. The horizontal gray stripes denote qualitatively ranges of t where the B atoms exhibit NM, MV and LM behavior. An interesting feature of the phase diagram is the appearance of a large FM region ($S_g \neq 0$) enclosed by the dashed line. The RKKY region at large t and large x originates from the B atoms which become magnetic. For the non FM configurations and for $x<5$ the Kondo (RKKY) correlations of the A atoms dominate at small (large) t, in analogy with the $x=0$ case. On the other hand, for $x=5$ the local Kondo correlations of the single

A atom at low t dominate over the f-f correlations between the A-B and B-B pairs. For the uniform (x=6) MV case we include only results in the large t regime, where the MV atoms acquire LM's which couple antiferromagnetically. Overall, the RKKY interactions prevail for any concentration when t is comparable or larger than the hybridization V.

8.4 Conclusions

Recent advances in STM experiments have made it possible to study the electronic and magnetic properties of strongly correlated electrons in nanoscopic and mesoscopic systems. There are two main differences between nanosized clusters and the infinite lattice. First, the discrete energy levels of the conduction band states introduce a new low-energy scale, i.e., the average energy level spacing Δ. This new energy scale that competes with the spin gap can effect the low-temperature behavior of the system. Second, the results depend on the parity of the total number of electrons. If N_{tot} is odd, the ground state is doubly degenerate.

We have carried out exact diagonalization calculations which reveal that the: (1) energy spacing; (2) parity of the number of electrons; and (3) disorder, give rise to a novel tuning of the magnetic behavior of a *dense* Kondo nanocluster. This interesting and important tuning can drive the nanocluster from the Kondo to the RKKY regime, i.e. a tunable "Doniach" phase diagram in nanoclusters. We have employed the criterion of comparing the exact *non local* versus *local* spin correlation functions to determine if the nanocluster lies in the RKKY versus Kondo regime. For weak hybridization, where the spin gap is smaller than Δ, both the low-temperature local f susceptibility and specific heat exhibit an exponential activation behavior associated with the spin gap. In contrast in the large hybridization limit, Δ is smaller than the spin gap, the physics become local and the exponential activation behavior disappears. The interplay of Δ and disorder produces a rich structure zero-temperature alloy phase diagram, where regions with prevailing Kondo or RKKY correlations alternate with domains of FM order. The distribution of local T_K and RKKY interactions depends strongly on the local environment and are overall *enhanced* by disorder, in contrast to the hypothesis of single-impurity based "Kondo disorder" models for extended systems. We believe that the conclusions of our calculations should be relevant to experimental realizations of small clusters and quantum dots. For example, the recent experiments[21] of magnetic clusters on single-walled carbon nanotubes of varying size provide much flexibility for investigating the interplay of Kondo and RKKY effects at different energy scales.

Acknowledgments

The research at California State University Northridge was supported through NSF under Grant Nos. DMR-0097187, NASA under grant No. NCC5-513, and the Keck and Parsons Foundations grants. The calculations were performed on the the CSUN Massively Parallel Computer Platform supported through NSF under Grand No. DMR-0011656. We acknowledge useful discussions with P. Fulde, P. Schlottmann, P. Riseborough, A.H. Castro Neto, P.Cornaglia and C. Balseiro.

References

[1] A.C.Hewson, *The Kondo Problem to Heavy Fermions*, Cambridge Press, New York, 1993.

[2] P.W.Anderson, Phys.Rev. **124** ,41 (1961); J.Kondo, Progr. Theor. Phys. **32** , 37 (1964).

[3] C.D.Bredl, S.Horn, F.Steglich, B.Luthi and R.M.Martin, Phys. Rev. Lett. **52**, 1982 (1984).

[4] H.Tsunetsugu, M. Sigrist and K.Ueda, Rev.Mod.Phys. **69** , 809 (1997).

[5] E.Miranda, V.Dobrosavljevic and G.Kotliar, Phys. Rev. Lett. **78**, 290 (1997).

[6] A.H.Castro Neto and B.A. Jones, Phys.Rev. B **62**, 14975 (2000).

[7] G.R.Stewart, Rev. Mod. Phys. **73**, 797 (2001).

[8] P.Schlottmann, Phys. Rev. B **65**, 174407 (2002).

[9] J. Lleweilun Smith and and Q. Si, Phys. Rev. B **61**, 5184 (2003).

[10] S.Doniach, Physica B **91**, 231 (1977).

[11] A. Schröder, G. Aeppli, R. Coldea, M. Adams, O. Stockert, H.V. Löhneysen, E. Bucher, R. Ramazashvili and P. Coleman, Nature **407**, 351 (2000).

[12] P.S.Riseborough, Phys.Rev.B **45**, 13984 (1992).

[13] O.O.Bernal, D.E. Maclaughlin, H.G.Lukefahr and B.Andraka, Phys. Rev. Lett. **75**, 2023 (1995).

[14] T.M.Rice and K.Ueda, Phys. Rev. Lett. **55**, 995 (1985).

[15] N.Read, D.M.Newns, and S.Doniach, Phys. Rev. B**30**, 3841 (1984); S.Burdin, A.Georges, and D.R.Grempel, Phys. Rev. Lett. **85**, 1048 (2000).

[16] S.Capponi and F.F.Assaad, Phys. Rev. B **63**,155114 (2001).

[17] E.Miranda and V. Dobrosavljevic, Phys. Rev. Lett. **86** 264 (2001).

[18] D.G.Gordon, H.Shtrikman, D. Mahalu, D.A. Magder, U.Meirav, and M.A.Kaster, Nature (London) **391**, 156 (1998).

[19] W.B. Thimm, J. Kroha, and J.V. Delft, Phys. Rev. Lett. **82**, 2143 (1999).

[20] H.C. Manoharan, C.P. Lutz, and D.M. Eigler, Nature (London) **403**, 512 (2000).

[21] T. Odom, J.L. Huang, C. Li Cheung, and C. M. Lieber, Science **290**, 1549 (2000) and references therein.

[22] H.Hu, G.M. Zhang and L. Yu, Phys. Rev. Lett. **86**, 5558 (2001).

[23] P.S. Cornaglia and C.A. Balseiro, Phys.Rev. B **66**,115303 (2002).

[24] P. Simon and I. Affleck, Phys.Rev.Lett **89**,206602,(2002).

[25] P. Schlottmann, Phys.Rev.B **65**, 024431 (2001).

[26] G.M.Pastor, R.Hirsch and B.Mühlschlegel, Phys.Rev.Lett. **72**, 3879 (1994).

[27] B.Coqblin, C. Lacroix, M.S. Gusmao and J.R. Iglesias, Phys.Rev. B **67**,064417(2003).

[28] K.Haule, J.Bonca and P. Prelovsek, Phys.Rev.B **61**, 2482 (2000).

[29] D.Meyer and W.Nolting, Phys.Rev.B **62**, 5657 (2000); D.Meyer, Solid State Comm. **121**, 565 (2002).

[30] B.K.Rao and P.Jena , Phys.Rev.Lett. **89**, 185504 (2002).

[31] P.Fulde, private communication, and P.Fulde, "Electron Correlations in Molecules and Solids", 3rd edition, Springer-Berlin-(1995).

[32] K.Yosida, Phys.Rev. **147**, 223 (1966).

[33] We also employed a second approach, $k_B T'_K = \mu_i B_c$, where B_c is the critical local external magnetic field necessary to break up the singlet bound state [23]. Comparative results of the methods will be presented elsewhere.

DENSITY FUNCTIONAL CALCULATIONS NEAR FERROMAGNETIC QUANTUM CRITICAL POINTS

I.I. Mazin, D.J. Singh and A. Aguayo

Center for Computational Materials Science
Naval Research Laboratory
Washington, DC 20375

Abstract We discuss the application of the density functional theory in the local density approximation (LDA) near a ferromagnetic quantum critical point. The LDA fails to describe the critical fluctuations in this regime. This provides a fingerprint of a materials near ferromagnetic quantum critical points: overestimation of the tendency to magnetism in the local density approximation. This is in contrast to the typical, but not universal, tendency of the LDA to underestimate the tendency to magnetism in strongly Hubbard correlated materials. We propose a method for correcting the local density calculations by including critical spin fluctuations. This is based on (1) Landau expansion for the free energy, evaluated within the LDA, (2) lowest order expansion of the RPA susceptibility in LDA and (3) extraction of the amplitude of the relevant (critical) fluctuations by applying the fluctuation-dissipation theorem to the difference between a quantum-critical system and a reference system removed from the quantum critical point. We illustrate some of the aspects of this by the cases of Ni_3Al and Ni_3Ga, which are very similar metals on opposite sides of a ferromagnetic quantum critical point. LDA calculations predict that Ni_3Ga is the more magnetic system, but we find that due to differences in the band structure, fluctuation effects are larger in Ni_3Ga, explaining the fact that experimentally it is the less magnetic of the two materials.

Keywords: quantum criticality, magnetism, density functional theory, first-principles calculation.

9.1 Introduction

Recent low temperature experiments on clean materials near ferromagnetic quantum critical points (FQCP) have revealed a remarkable range of unusual properties, including non-Fermi liquid scalings over a

S. Halilov (ed.), Physics of Spin in Solids: Materials, Methods and Applications, 139–154.

large phase space, unusual transport, and novel quantum ground states, particularly coexisting ferromagnetism and superconductivity in some materials. Although criticality usually implies a certain universality, present experiments show considerable material dependent aspects that are not well understood, [1] *e.g.* the differences between UGe_2 and URhGe [24, 25] and $ZrZn_2$, [4] which both show coexisting ferromagnetism and superconductivity but very different phase diagrams, in contrast to MnSi, where very clean samples show no hint of superconductivity around the QCP, possibly because of the lack of the inversion symmetry. [5]

Moreover, by far not every magnetic material can be driven to a QCP by pressure or by other means of supressing ferromagnetism. Typically, the transition becomes first order as the Curie temperature, T_C is depressed. If this happens too far away from the fluctuation dominated regime, nothing interesting is seen. Also, more pedestrian effects are often important. For example, impurities or other defects can lead to scattering that smears out the quantum critical region.

9.2 The LDA Description Near a FQCP

One of the fingerprints of a FQCP, maybe the most universal one, is a substantial overestimation of the tendency to magnetism in conventional density functional theory (DFT) calculations, such as within the local density approximation (LDA). Generally, approaches based on density functional theory (DFT) are successful in accounting for material dependence in cases where sufficiently accurate approximations exist. Density functional theory is in principle an exact ground state theory. It should, therefore, correctly describe the spin density of magnetic systems. This is usually the case in actual state of the art density functional calculations. However, common approximations to the exact density functional theory, such as the LDA, may miss important physics and indeed fail to describe some materials. A well know example is in strongly Hubbard correlated systems, where the LDA treats the correlations in an orbitally averaged mean field way and often underestimates the tendency towards magnetism.

Overestimates of magnetic tendencies, especially in the LDA, are considerably less common, the exceptions being materials near magnetic quantum critical points (QCP); here the error comes from neglect of low energy quantum spin fluctuations. In particular, the LDA is parameterized based on the uniform electron gas at densities typical for atoms and solids. However, the uniform electron gas at these densities is stiff against magnetic degrees of freedom and far from magnetic

Table 1. Some materials near a FQCP that we have investigated by LDA calculations. Type 1 materials are ferromagnetic both in the calculations and in experiment; magnetic moments in μ_B per formula unit are given. Type 2 are ferromagnetic only in the calculations (calculated moments given, and type 3 are paramagnetic (susceptibility in 10^{-4}emu/mol is given). The references are to the LDA calculations.

Material	$ZrZn_2$	Ni_3Al	Sc_3In	FeAl	Ni_3Ga
Type	1	1	1	2	2
Calc/Exp	0.72/0.17	0.71/0.23	1.05/0.20	0.80	0.79
Ref.	[16]	[17]	[18]	[19]	[17]

Material	$Sr_3Ru_2O_7$	$SrRhO_3$	$Na_{0.5}CoO_2$	Pd
Type	2	2	2	3
Calc/Exp	0.80	0.9	0.5	11.6/6.8
Ref.	[8]	[20]	[10]	[21]

QCP's. Thus, although the LDA is exact for the uniform electron gas, and therefore does include all fluctuation effects there, its description of magnetic ground states in solids and molecules is mean field like. This leads to problems such as the incorrect description of singlet states in molecules with magnetic ions as well as errors in solids when spin fluctuation effects beyond the mean field are important. In solids near a QCP, the result is an overestimate of the magnetic moments and tendency toward magnetism (*i.e.* misplacement of the position of the critical point) due to neglect of the quantum critical fluctuations. [6, 7] Examples include three types of materials: paramagnets that are ferromagnetic in the LDA, ferromagnets where the equilibrium magnetic moment is substantially overestimated in the LDA, and paramagnets where the paramagnetic susceptibility is substantially overestimated.

We list examples of materials in all three categories in Table 1. At least two of these are cases where a large deviation between the LDA and experimental magnetic properties were noted, followed by transport measurements that suggest a nearby ferromagnetic quantum critical point. In particular, in $Sr_3Ru_2O_7$, LDA calculations with the experimental crystal structure found a sizeable moment, [8] while experimentally the material was known to be a paramagnetic metal. Grigera and coworkers then showed that $Sr_3Ru_2O_7$ has a metamagnetic quantum critical point at moderate field. [9] Pd metal provides another example: the calculated LDA magnetic susceptibility is nearly twice larger that the experimental one. Correspondingly, Nicklas *et al*[11] found a FQCP in the $Pd_{1-x}Ni_x$ system at $x = 0.026$, where the transport properties become non-Fermi liquid.

We emphasize that substantial overestimates of the tendency of metals towards ferromagnetism within the LDA is a rare occurance, and propose that it be used as an indicator of critical fluctuations in a material. However, for this to be an effective screen, competing states, like antiferromagnetism need to be ruled out in each material. An interesting case study is LiV_2O_4, which is a paramagnetic metal and occurs in the cubic spinel structure. Remarkably, it was discovered by Kondo and co-workers that this material behaves at low temperature like a heavy fermion metal. [12] LDA calculations showed that the material is unstable against ferromagnetism with a sizeable moment. [14, 15, 13] But calculations also show that the interactions are antiferromagnetic, and as a result it is more unstable against antiferromagnetism, which however is frustrated on the spinel lattice. While LiV_2O_4 may be near an antiferromagnetic QCP, it is not a material near an FQCP.

9.3 "Beyond-LDA" Critical Fluctuations

A popular way to add quantum or termal fluctuation to a mean-field type theory is *via* fluctuation corrections to Ginzburg-Landau expansion of the free energy. For a detailed discussion we refer the reader to the book of Moriya [22] and the review article of Shimizu [23]. In short, one writes the free energy (or the magnetic field) as a function of the ferromagnetic magnetization, M,

$$E_{static}(M) = a_0 + \sum_{n\geq 1} \frac{1}{2n} a_{2n} M^{2n}, \tag{1}$$

$$H_{static}(M) = \sum_{n\geq 1} a_{2n} M^{2n-1} \tag{2}$$

(obviously, a_2 gives the inverse spin susceptibility without fluctuations), and then assume Gaussian zero-point fluctuations of an r.m.s. magnitude ξ for each of the d components of the magnetic moment (for a 3D isotropic material like Pd, $d = 3$). After averaging over the spin fluctuations, one obtains a fluctuation-corrected functional. The general expression can be written in the following compact form:

$$H(M) = \sum_{n\geq 1} \tilde{a}_{2n} M^{2n-1}$$

$$\tilde{a}_{2n} = \sum_{i\geq 0} C^{n-1}_{n+i-1} a_{2(n+i)} \xi^{2i} \Pi^{n+i-1}_{k=n} (1 + \frac{2k}{d}). \tag{3}$$

For instance,

$$\tilde{a}_2 = a_2 + \frac{5}{3} a_4 \xi^2 + \frac{35}{9} a_6 \xi^4 + \frac{35}{3} a_8 \xi^6 ...$$

$$
\begin{aligned}
\tilde{a}_4 &= a_4 + \frac{14}{3} a_6 \xi^2 + 21 a_8 \xi^6 ... \\
&...
\end{aligned} \tag{4}
$$

The unrenormalized coefficients can be taken from fixed spin momen LDA calculations, in which case ξ becomes the amplitude of those fluctuations only, which are not taken into account in LDA (as mentioned, LDA includes some quantum fluctuation, specifically short-range fluctuations present in the interacting uniform electron gas). In principle, one can estimate ξ from the fluctuation-dissipation theorem, which states that (see, e.g., Refs. [24, 25])

$$
\xi^2 = \frac{4\hbar}{\Omega} \int d^3 q \int \frac{d\omega}{2\pi} \frac{1}{2} \operatorname{Im} \chi(\mathbf{q}, \omega), \tag{5}
$$

where $\chi(\mathbf{q}, \omega)$ is the magnetic susceptibility and Ω is the Brillouin zone volume. It is customary to approximate $\chi(\mathbf{q}, \omega)$ by its small q, small ω expansion [24, 25]:

$$
\chi_0(\mathbf{q}, \omega) = N(E_F) - a q^2 + i b \omega / q \tag{6}
$$

$$
\chi^{-1}(\mathbf{q}, \omega) = \chi_0^{-1}(\mathbf{q}, \omega) - I, \tag{7}
$$

With the expansion (6) the integrations can be performed analytically, and the final result reads:

$$
\xi^2 = \frac{b v_F^2 N(E_F)^2}{2 a^2 \Omega} [Q^4 \ln(1 + Q^{-4}) + \ln(1 + Q^4)]. \tag{8}
$$

where $Q = q_c \sqrt{a / b v_F}$, and q_c is the cutoff parameter for momentum integration in Eqn. 5 (the frequency integration at a given q is usually assumed to be cut off at $\omega = v_F q$).

To proceed along these lines one needs to find a way to calculate the crucial parameters of the expansion (6). It was suggested by Moriya [22] that these can be expressed as certain integrals over the Fermi surface, by expanding the RPA expression for χ_0. Below, we offer a derivation equivalent to that of Moriya, but rendering the results in more computable form. We start with the RPA expressions for the real and imaginary parts of χ_0 :

$$
\operatorname{Re} \chi_0(\mathbf{q}, 0) = \sum_{\mathbf{k}} [f(E_{\mathbf{k}}) - f(E_{\mathbf{k}+\mathbf{q}})] (E_{\mathbf{k}+\mathbf{q}} - E_{\mathbf{k}})^{-1} \tag{9}
$$

$$
\operatorname{Im} \chi_0(\mathbf{q}, \omega) = \sum_{\mathbf{k}} [f(E_{\mathbf{k}}) - f(E_{\mathbf{k}+\mathbf{q}})] \delta(E_{\mathbf{k}+\mathbf{q}} - E_{\mathbf{k}} - \omega), \tag{10}
$$

where $f(E)$ is the Fermi function, $-\frac{df(E)}{dE} = \delta(E - E_F)$. Expanding Eqn. 9 in $\Delta = E_{\mathbf{k}+\mathbf{q}} - E_{\mathbf{k}} = \mathbf{v}_{\mathbf{k}} \cdot \mathbf{q} + \frac{1}{2} \sum_{\alpha\beta} \mu_{\mathbf{k}}^{\alpha\beta} q_\alpha q_\beta + ...$, we get to second order in q

$$\mathrm{Re}\,\chi_0(\mathbf{q},\mathbf{0}) = N(E_F) + \sum_{\mathbf{k}}[\frac{1}{2}\left(\frac{d\delta(\varepsilon_{\mathbf{k}}-E_F)}{dE_F}\right)(\mathbf{v}_{\mathbf{k}}\cdot\mathbf{q}+\frac{1}{2}\sum_{\alpha,\beta}\mu_{\mathbf{k}}^{\alpha\beta}q_\alpha q_\beta)$$
$$+\frac{1}{6}\left(\frac{d^2\delta(\varepsilon_{\mathbf{k}}-E_F)}{dE_F^2}\right)(\mathbf{v}_{\mathbf{k}}\cdot\mathbf{q})^2].$$

The odd powers of $\mathbf{v}_{\mathbf{k}}$ cancel out and we get $(\alpha,\beta = x,y,z)$

$$\mathrm{Re}\,\chi_0(\mathbf{q}) =$$
$$N(E_F) + \sum_{\alpha,\beta}\frac{q_\alpha q_\beta}{4}\frac{d\left\langle N(E_F)\mu^{\alpha\beta}\right\rangle}{dE_F} + \sum_{\alpha,\beta}\frac{q_\alpha q_\beta}{6}\frac{d^2\left\langle N(E_F)v_\alpha v_\beta\right\rangle}{dE_F^2}$$
$$= N(E_F) + \frac{q^2}{4}\frac{d\left\langle N(E_F)\mu_{xx}\right\rangle}{dE_F} + \frac{q^2}{6}\frac{d^2\left\langle N(E_F)v_x^2\right\rangle}{dE_F^2},$$

where $v_x^2 = v_y^2 = v_z^2$, $\mu_{xx} = \mu_{yy} = \mu_{zz}$. The last equality assumes cubic symmetry; generalization to a lower symmetry is trivial. Using the following relation,

$$\sum_{\mathbf{k}}\nabla_{\mathbf{k}}F(\varepsilon_{\mathbf{k}}) = \sum_{\mathbf{k}}\frac{dF(\varepsilon_{\mathbf{k}})}{d\varepsilon_{\mathbf{k}}}\nabla_{\mathbf{k}}\cdot\varepsilon_{\mathbf{k}} = \sum_{\mathbf{k}}\frac{dF(\varepsilon_{\mathbf{k}})}{d\varepsilon_{\mathbf{k}}}\mathbf{v}_{\mathbf{k}},$$

one can prove that

$$\frac{d^2\left\langle N(E_F)v_x^2\right\rangle}{dE_F^2} = -\frac{d\left\langle N(E_F)\mu_{xx}\right\rangle}{dE_F}. \tag{11}$$

Therefore

$$\mathrm{Re}\,\chi_0(\mathbf{q}) = N(E_F) - \frac{q^2}{12}\frac{d^2\left\langle N(E_F)v_x^2\right\rangle}{dE_F^2} \tag{12}$$

Similarly, for Eqn. 10 one has

$$\mathrm{Im}\,\chi_0(\mathbf{q},\omega) = \sum_{\mathbf{k}}\left[\left(-\frac{df(\varepsilon)}{d\varepsilon}\right)\omega\delta(\mathbf{v}_{\mathbf{k}}\cdot\mathbf{q}-\omega)\right] \tag{13}$$

After averaging over the directions of $\mathbf{q}$, this becomes, for small ω,

$$\mathrm{Im}\,\chi_0(q,\omega) = \frac{\omega}{2}\sum_{\mathbf{k}}\frac{\delta(\varepsilon_{\mathbf{k}})}{v_{\mathbf{k}}q}\theta(v_{\mathbf{k}}q-\omega) = \frac{\omega}{2q}\left\langle N(E_F)v^{-1}\right\rangle$$
$$v = \sqrt{v_x^2+v_y^2+v_z^2}. \tag{14}$$

Although in real materials the Fermi velocity is obviously different along different directions, it is still a reasonable approximation to introduce an average v_F. Then the above formulae reduce all parameters needed for estimating the *r.m.s.* amplitude of the spin fluctuations to four integrals over the Fermi surface, specifically, the density of states, $N(E_F)$, $a = \frac{1}{12}\frac{d^2\langle N(E_F)v_x^2\rangle}{dE_F^2}$, $b = \frac{1}{2}\langle N(E_F)v^{-1}\rangle$ and $v_F = \sqrt{3\frac{\langle N(E_F)v_x^2\rangle}{N(E_F)}}$.

The physical meaning of these parameters is as follows. a defines the rate at which the static susceptibility $\chi(q,0)$ falls away from the zone center, *i.e.* the extent to which the tendency to ferromagnetism is stronger than that to antiferromagnetism. This translates into the phase space in the Brillouin zone where the spin fluctuations are important. b controls the dynamic effects in spin susceptibility.

Note that the cutoff parameter q_c remains the only undefined quantity in this formalism. One obvious choice is $q_c = \sqrt{N(E_F)/a}$, because for larger q the approximation (6) gives unphysical negative values for the static susceptibility. On the other hand, one may argue that q_c should reflect mainly the geometry of the Fermi surface and thus not depend on a at all. We will come back to this issue later in this paper and will propose an approach that avoids using q_c whatsoever.

9.4 Ni_3Al and Ni_3Ga

Here we use the closely related compounds Ni_3Al and Ni_3Ga to illustrate some of the above ideas. Further details may be found in Ref. [17]. These have the ideal cubic Cu_3Au $cP4$ structure, with very similar lattice constants, $a = 3.568$ Å and $a = 3.576$ Å, respectively, and have been extensively studied by various experimental techniques. Ni_3Al is a weak itinerant ferromagnet, $T_c = 41.5$ K and magnetization, M=0.23 μ_B/cell (0.077 μ_B/Ni atom) [26] with a QCP under pressure at P_c=8.1 GPa, [27] while Ni_3Ga is a strongly renormalized paramagnet. [28] Further, it was recently reported that Ni_3Al shows non-Fermi liquid transport over a large range of P and T range down to very low T. [29]

Previous LDA calculations showed that the magnetic tendency of both materials is overestimated within the LDA, and that Ni_3Ga is incorrectly predicted to be a ferromagnet. [30–35] Moreover, in the LDA the tendency to magnetism is stronger in Ni_3Ga than Ni_3Al, *opposite to the experimental trend.* This poses an additional challenge to any theory striving to describe the material dependent aspects of quantum criticality. The two materials are expected to be very similar electronically (the small difference between the two is due to relativistic effects associated with Ga in Ni_3Ga). Thus these two very similar metals offer a very useful and sensitive benchmark for theoretical approaches. We use this to

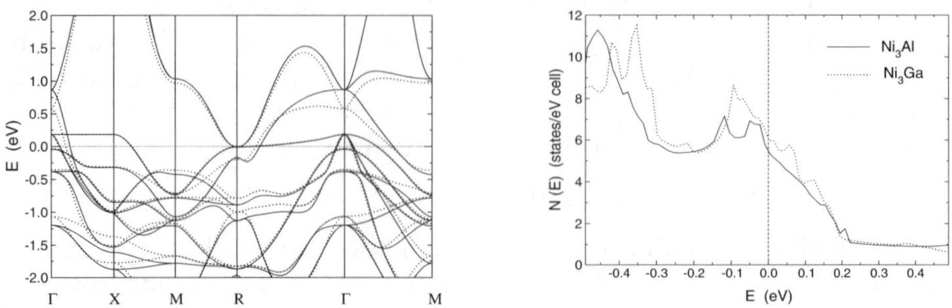

Figure 1. Calculated LDA band structure (left) and density of states (right) per f.u. for non-spin-polarized Ni_3Al (solid lines) and Ni_3Ga (dotted lines). E_F is at 0 eV.

test an approach based on the fluctuation dissipation theorem applied to the LDA band structures with an ansatz for the cut-off q_c. We find that this approach corrects the ordering of the magnetic tendencies of the materials, and gives the right ground states at ambient pressure as well as a reasonable value of P_c for Ni_3Al.

The LDA calculations were done using the general potential linearized augmented planewave (LAPW) method with local orbital extensions [36, 37, 39] as decribed in Ref. [17], with the exchange-correlation functional of Hedin and Lundqvist with the von Barth-Hedin spin scaling [40, 41]. The LDA electronic structure is given in Fig. 1 and Table 2, while results of fixed spin moment calculations of the magnetic properties at the experimental lattice parameters and under hydrostatic compression are given in Figs. 2 and 3. The two compounds are very similar in both electronic and magnetic properties, the main apparent difference being the higher equilibrium moment of Ni_3Ga (0.79 μ_B/f.u. *vs.* 0.71 μ_B/f.u.), in agreement with other full potential calculations. [34, 35]

The propensity towards magnetism may be described in terms of the Stoner criterion, $IN(E_F)$, where I is the so-called Stoner parameter, which derives from Hund's rule coupling on the atoms. For finite magnetizations, the so-called extended Stoner model [42], states that, to the second order in the spin density, the magnetic stabilization energy is given by

$$\Delta E = M^2[\int_0^M m\ dm/2\tilde{N}(m) - I/4], \qquad (15)$$

where $\tilde{N}(M)$ is the density of states averaged over the exchange splitting corresponding to the magnetization M. Fitting the fixed spin moment results to this expression, we find $I_{Al} = 0.385$ eV and $I_{Ga} = 0.363$ eV. These gives $IN(E_F)$ =1.21 and $IN(E_F) = 1.25$ for Ni_3Al and Ni_3Ga,

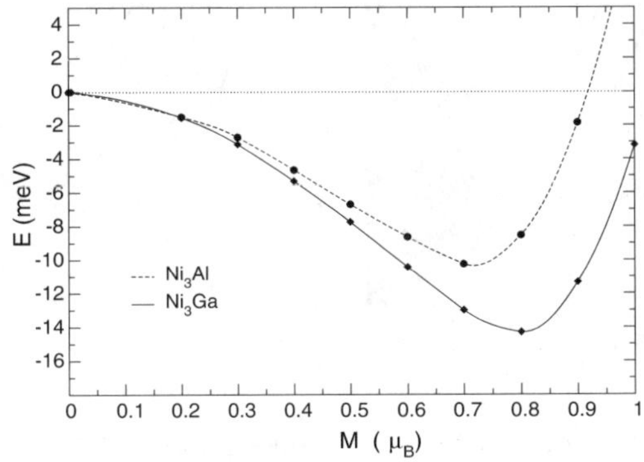

Figure 2. Energy *vs.* fixed spin moment for Ni_3Al and Ni_3Ga at the experimental lattice parameters. The energy zero is set to the non-spin-polarized value.

respectively. Both numbers are larger than unity, corresponding to a ferromagnetic instability, and the value for Ni_3Ga is larger than that for Ni_3Al. Importantly, the difference comes from the density of states, since $I_{Al} > I_{Ga}$. In both compounds, magnetism is suppressed by compression, with an LDA critical point at a value $\delta a/a \sim$ -0.05 – -0.06. In Ni_3Al, the critical point at $\delta a/a$ =-0.058 corresponds to the pressure of P_c =50 GPa, [43] which is much higher than the experimental value. It is interesting that, as in $ZrZn_2$ [16], the exchange splitting is very strongly **k**-dependent; for instance, in Ni_3Al at some points it is as small as 40 meV/μ_B near the Fermi level, while at the others (of pure Ni d character) it is close to 220 meV/μ_B.

Notwithstanding the general similarity of the two compounds, there is one important difference near the Fermi level, specifically, the light band crossing the Fermi level in the middle of the Γ-M or Γ-X directions is steeper in Ni_3Al (Fig. 1). This, in turn, leads to smaller density of states. This comes from a different position of the top of this band at the Γ point, 0.56 eV in Ni_3Ga and 0.85 eV in Ni_3Al. The corresponding electronic state is a mixture of Ni *p* and Al (Ga) *p* states, and is the only state near the Fermi level with substantial Al (Ga) content. Due to relativistic effects, the Ga *p* level is lower than the Al *p* level and this leads to the difference in the position of the corresponding hybridized state. Note that this is a purely scalar relativistic effect. Including spin orbit does not produce any further discernible difference.

Returning to magnetism, the fixed spin moment calculations provide the energy E as a function of the magnetization M (Fig. 2). One can

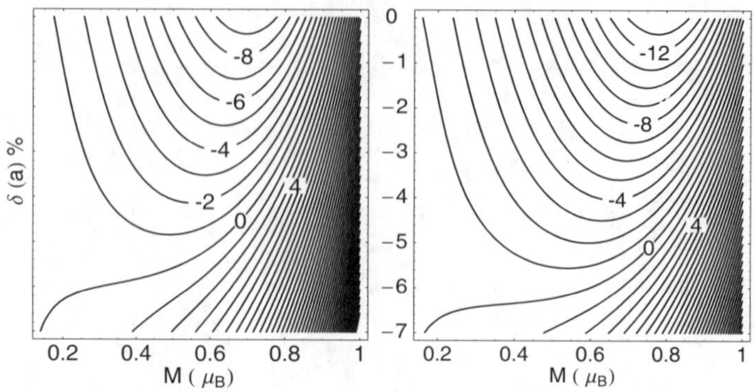

Figure 3. FSM calculations under hydrostatic pressures. Magnetic energy, defined as the energy relative to the non-spin-polarized result at the same volume, as a function of the moment and linear compression. Left and right panels correspond to Ni_3Al and Ni_3Ga, respectively.

Table 2. Magnetic energy (see text), magnetic moment in μ_B/cell and $N(E_F)$ in eV^{-1} for Ni_3Al and Ni_3Ga on a per spin per formula unit basis.

	$\|\Delta E\|$ (meV)	M (calc.)	M(expt.)	$N(E_F)$
Ni_3Al	10.3	0.71	0.23	3.2
Ni_3Ga	14.3	0.79	0.00	3.4

write a Landau expansion for $E(M)$ as in Eqn. 1, which may then be treated as a mean field expression adding the effects of spin fluctuations. [23]

Treating this as a mean field expression and adding the effects of spin fluctuations [23] leads to renormalization of the expansion coefficients. The renormalized coefficients $\tilde{a}_i$ are written as power series in the averaged square of the magnetic moment fluctuations beyond the LDA, ξ^2 as in Eqn. 3. ξ may then be estimated by requiring that the corrected Landau functional reproduces the experimental magnetic moment (for Ni_3Al) or experimental magnetic susceptibility (for Ni_3Ga). The "experimental" ξ's obtained in this manner are are 0.47 and 0.55, respectively, which implies that spin fluctuation effects must be stronger in Ni_3Ga than in Ni_3Al.

A link can now be made between this fact and the electronic structures, using the formalism outlined in the previous section. As discussed, the cutoff parameter q_c is the least well defined quantity in this formalism. Furthermore, the fermiology of these compounds is very complicated: in the paramagnetic state, there are four Fermi surfaces, two small and two large (one open and one closed). In this situation, it is hardly possible to justify any simple prescription for q_c. Therefore, we chose a different route: we assume that q_c is the *same* for both materials, and choose a number which yields a good description of both the equilibrium moment in Ni_3Al and the paramagnetic susceptibility in Ni_3Ga, $q_c = 0.382\ a_0^{-1}$. Note that this is larger that the diameters of the small Fermi surfaces but smaller than the radius of the Brillouin zone, $\approx 0.5\ a_0^{-1}$.

To calculate the above quantities, especially a, we need accurate values of the velocities on a fine mesh. Numerical differentiation of energies within the tetrahedron method proved to be too noisy. Therefore we use the velocities obtained analytically as matrix elements of the momentum operator, computed within the *optic* program of the WIEN package. A bootstrap method, [44] as described in Ref. [21], was used to obtain stable values for a, b. We found for Ni_3Al, using as the energy unit Ry, the length unit Bohr, and the velocity unit Ry·Bohr, $a = 230$, $b = 210$, $v_F = 0.20$, and $\xi = 0.445\ \mu_B$. For Ni_3Ga $a = 140$, $b = 270$, $v_F = 0.19$, and $\xi = 0.556\ \mu_B$. Using the resulting values of ξ each compound we obtain a magnetic moment of $M = 0.3\ \mu_B$/cell for Ni_3Al and a paramagnetic state with the renormalized susceptibility $\chi(0,0) = 1/\tilde{a}_2 = 6.8 \times 10^{-5}$ emu/g for Ni_3Ga, thus correcting the incorrect ordering of the magnetic tendencies of these two compounds and reproducing extremely well the experimental numbers of $M = 0.23\ \mu_B$, $\chi(0,0) = 6.7 \times 10^{-5}$ emu/g, respectively. This qualitative behavior is due to the different coefficient a, *i.e.*, different q dependencies of $\chi_0(q,0)$ at small q, which relates to the phase space available for soft fluctuations.

Now we turn to the pressure dependence. The above results imply that beyond-LDA fluctuations are already larger than the moments themselves at $P = 0$. In this regime, we may assume that the size of the beyond-LDA fluctuations is only weakly pressure dependent. Then we can apply Eqn. 3 to the data shown in Fig. 3 using $\xi = 0.47$ as needed to match the $P = 0$ value of M. This yields a value P_c=10 GPa in quite good agreement with the experimental value, P_c=8.1 GPa. [27]

9.5 Towards a Fully First Principles Theory

The results for Ni_3Al and Ni_3Ga, discussed above, and in Ref. [17], show that an approach based on correction of the LDA using the fluctuation dissipation theorem has promise. However, the results hinge on an unknown cut-off, which serves the purpose of including fluctuations that are associated with the FQCP and are not included in the LDA, from those that are included in the LDA. While it is apparently possible to obtain useful results using reasonable ansatz for this cut-off, it would be much better to have a truly first principles theory, with no parameters. In order to construct such a theory, one should find a way of solving the double counting problem, *i.e* including in the correction only those fluctuations that are not already taken into acount at the LDA level. This amounts to subtracting from Eqn. 5 the fluctuations already included in the LDA. Since the LDA is known to work well for materials far from an FQCP, this means that the correction should be zero or close to it for the most materials.

We suggest that a consistent way to accomplish this is by introducing a "reference" susceptibility $\chi^{ref}(\mathbf{q},\omega)$ and subtracting it from $\chi(\mathbf{q},\omega)$:

$$\xi^2 = \frac{4\hbar}{\Omega}\int d^3q \int \frac{d\omega}{2\pi}\frac{1}{2}\,\mathrm{Im}[\chi(\mathbf{q},\omega) - \chi^{ref}(\mathbf{q},\omega)], \tag{16}$$

We shall use the same expansion6 for both $\chi(\mathbf{q},\omega)$ and $\chi^{ref}(\mathbf{q},\omega)$, to derive equivalent expansions

$$\chi^{-1}(\mathbf{q},\omega) = \chi_0^{-1}(0,0) - I + Aq^2 - iB\omega/q, \tag{17}$$

where $\chi_0^{-1}(0,0) = 1/N(E_F)$ (density of states per spin) is the bare (non-interacting) static uniform susceptibility, and the Stoner parameter I is only weakly dependent on $\mathbf{q}$ and ω. Note that $A = a/N^2$, $B = b/N^2$, where a and b are the coefficients introduced in Eq.6. We also introduce a notation, $\Delta = N(E_F)^{-1} - I$. As long the same functional form (17) is used for $\chi(\mathbf{q},\omega)$ and $\chi^{ref}(\mathbf{q},\omega)$, the condition for the convergence of the integral (16) is that the coefficients A and B, controlling the short-range and high frequency fluctuations are the same. Of course, the parameter Δ, defining the proximity to the QCP, is different in the reference system, which like the uniform electron gas upon which the LDA is based, should be far from any QCP (let us call Δ for the reference system Δ_0).

To calculate the integral ((16), we write it in the following form:

$$\xi^2 = \frac{4\hbar}{\Omega}\int d^3q \int \frac{d\omega}{2\pi}\frac{1}{2}\,\mathrm{Im}[\chi(\Delta,\mathbf{q},\omega) - \chi(\Delta_0,\mathbf{q},\omega)]. \tag{18}$$

For instance, $\chi(0,\mathbf{q},\omega)$ represents the susceptibility right at the FQCP. This diverges for $\mathbf{q}=0$, $\omega=0$. The derivation then proceeds as follows:

$$\int^{\omega_c} d\omega\,\mathrm{Im}[\chi(\Delta,q,\omega)] = \frac{q}{2B}\ln[\frac{(\Delta+Aq^2)^2 + B^2\omega_c^2/q^2}{(\Delta+Aq^2)^2}]. \tag{19}$$

Where we introduce the Landau cutoff frequency, $\omega_c = vq$ (here v is an average Fermi velocity) and the notation $\beta = Bv$. We will also need the following function:

$$\begin{aligned} F(\Delta,\beta,x) &= \int x^3 dx \ln[(\Delta+x^2)^2+\beta^2] \\ &= \frac{(\Delta+x^2)^2+\beta^2}{4}\{\ln[(\Delta+x^2)^2+\beta^2]-1\} \\ &- \frac{\Delta(\Delta+x^2)}{2}\{\ln[(\Delta+x^2)^2+\beta^2]-2\}+\beta\Delta\tan^{-1}\frac{\beta}{\Delta+x^2} \end{aligned}$$

Now

$$\begin{aligned} \xi^2 &= \frac{2}{\Omega A^2 B}\lim_{Q\to\infty}[F(\Delta,\beta,Q)-F(\Delta_0,\beta,Q) \\ &\quad -F(\Delta,0,Q)+F(\Delta_0,0,Q)-F(\Delta,\beta,0) \\ &\quad +F(\Delta_0,\beta,0)+F(\Delta,0,0)-F(\Delta_0,0,0)]. \end{aligned}$$

This is particularly easy to evaluate at $\Delta = 0$. The result is

$$\begin{aligned} &\Xi^2(\Delta_0) = \\ &\frac{2}{\Omega A^2 B}[\Delta_0\beta(\frac{\pi}{2}-\tan^{-1}\frac{\Delta_0}{\beta})+\frac{\beta^2-\Delta_0^2}{4}\ln\frac{\Delta_0^2+\beta^2}{\beta^2}+\frac{\Delta_0^2}{4}\ln\frac{\Delta_0^2}{\beta^2}] \\ &\Xi^2(S_0) = \frac{N^2 b v_F^2}{2\Omega a^2}[4S_0\tan^{-1}(S_0^{-1})+\ln(1+S_0^2)-S_0^2\ln(1+S_0^{-2})], \end{aligned}$$

where $S_0 = \Delta_0 N^2/bv_F$. Obviously, for arbitrary Δ the answer is simply

$$\xi^2 = \Xi^2(S_0) - \Xi^2(S). \tag{20}$$

Given that usually the reason for a quantum criticality is a large density of states, it makes sense to take the Stoner parameter for the reference system the same as for the system in question. The point is that the density of states is a highly non-local parameter (note that it involves a delta function integral in energy), which can hardly be discerned from local information about the charge density, while the Stoner parameter is a very local quantity associated with the exchange-correlation potential. The difference between Δ and Δ_0 then comes from the difference between $N = N(E_F)$ and the density of states, N_0, of the reference system.

One may think about several different ways for choosing N_0. One may be to take average $N(E)$ over the width of the valence band, $N_0 = n/t$, where n is the total number of states in the band and t is its width. One can also think about the density of states of the uniform electron gas with the same Stoner parameter. There may be other, more sophisticated prescriptions. Probably, the most practical approach will be found after several trial and error tests with real materials.

9.6 Summary and Open Questions

The failure of the usual approximations to density functional theory, for example, the LDA, to describe the magnetic properties of materials near ferromagnetic quantum critical points is associated with renormalization due to critical fluctuations. It is pointed out that since such fluctuations are invariably antagonistic to ferromagnetic ordering, deviations between experiment and LDA calculations in which the LDA is overly ferromagnetic can be a useful screen for materials near FQCPs. These errors in the LDA can be corrected using a phenomenalogical Landau function approach with the fluctuation amplitude as a parameter. However, there is hope that this parameter can be obtained from the electronic structure via the fluctuation dissipation theorem and a suitable reference system. The key remaining challenges in our view are to define the reference system to be used, and to use calculations to determine the usefulness of this approach for real materials near a critical point.

Acknowledgments

We are grateful for helpful conversations with S.V. Halilov, G. Lonzarich and S. Saxena. Work at the Naval Research Laboratory is supported by the Office of Naval Research.

References

[1] R.B. Laughlin, G.G. Lonzarich, P. Monthoux and D. Pines, Adv. Phys. **50**, 361 (2001).

[2] S.S. Saxena, P. Agarwal, K. Ahilan, F.M. Grosche, R.K.W. Haselwimmer, M.J. Steiner, E. Pugh, I.R. Walker, S.R. Julian, P. Monthoux, G.G. Lonzarich, A. Huxley, I. Sheikin, D. Braithwaite, and J. Flouquet, Nature **406**, 587 (2000).

[3] D. Aoki, A. Huxley, E. Ressouche, D. Braithwaite, J. Flouquet, J.P. Brison, E. Lhotel, and C. Paulsen, Nature **413**, 613 (2001).

[4] C. Pfleiderer, M. Uhlarz, S.M. Hayden, R. Vollmer, H. von Lohneysen, N.R. Bernhoeft, and G.G. Lonzarich, Nature **412**, 58 (2001).

[5] C.P. Pfleiderer, S.R. Julian and G.G. Lonzarich, Nature **414**, 427 (2001).

[6] H. Yamada, K. Fukamichi and T. Goto, Phys. Rev. B **65**, 024413 (2001).

[7] A.J. Millis, A.J. Schofield, G.G. Lonzarich and S.A. Grigera, Phys. Rev. Lett. **88**, 217204 (2002).

[8] D.J. Singh and I.I. Mazin, Phys. Rev. B **63**, 165101 (2001).

[9] S.A. Grigera, R.S. Perry, A.J. Schofield, M. Chiao, S.R. Julian, G.G. Lonzarich, S.I. Ikeda, Y. Maeno, A.J. Millis, and A.P. Mackenzie, Science **294**, 329 (2001).

[10] D.J. Singh, Phys. Rev. B **61**, 13397 (2000); D.J. Singh, Phys. Rev. B **68**, 020503 (2003).

[11] M. Nicklas, M. Brando, G. Knebel, F. Mayr, W. Trinkl, and A. Loidl, Phys. Rev. Lett. **82**, 4268 (1999).

[12] S. Kondo, D.C. Johnston, C.A. Swenson, F. Borsa, A.V. Mahajan, L.L. Miller, T. Gu, A.I. Goldman, M.B. Maple, D.A. Gajewski, E.J. Freeman, N.R. Dilley, R.P. Dickey, J. Merrin, K. Kojima, G.M. Luke, Y.J. Uemura, O. Chmaissem and J.D. Jorgensen, Phys. Rev. Lett. **78**, 3729 (1997).

[13] D.J. Singh, P. Blaha, K. Schwarz, and I. I. Mazin, Phys. Rev. B **60**, 16359 (1999).

[14] V. Eyert, K.H. Hock, S. Horn, A. Loidl and P.S. Riseborough, Europhysics Lett. **46**, 762 (1999).

[15] V.I. Anisimov, M.A. Korotin, M. Zölfl, T. Pruschke, K. Le Hur and T.M. Rice, Phys. Rev. Lett. **83**, 364 (1999).

[16] D.J. Singh and I.I. Mazin, Phys. Rev. B **69**, 020402 (2004).

[17] A. Aguayo, I.I. Mazin, and D.J. Singh, cond-mat/0310629.

[18] A. Aguayo and D.J. Singh, Phys. Rev. B **66**, 020401 (2002).

[19] A.G. Petukhov, I.I.Mazin, L. Chioncel and A. I. Lichtenstein, Phys. Rev. B **67**, 153106 (2003).

[20] D.J. Singh, Phys. Rev. B **67**, 054507 (2003).

[21] P. Larson, I.I. Mazin and D.J. Singh, cond-mat/0305407 (Phys. Rev. B, January 2004, in press).

[22] T. Moriya, *Spin fluctuations in itinerant electron magnetism* (Berlin, Springer, 1985).

[23] M. Shimizu, Rep. Prog. Phys. **44**, 329 (1981).

[24] A.Z. Solontsov and D. Wagner, Phys. Rev. **B51**, 12410 (1995).

[25] S.N. Kaul, J. Phys. Cond. Mat. **11**, 7597 (1999).

[26] F.R. de Boer, C.J. Schinkel, J. Biesterbos, and S. Proost, J. Appl. Phys. **40**, 1049 (1969).

[27] P.G. Niklowitz, F. Beckers, N. Bernhoeft, D. Braithwaite, G. Knebel, B. Salce, J. Thomasson, J. Floquet and G.G. Lonzarich (unpublished); presented at Conference on Quantum Complexities in Condensed Matter, 2003.

[28] S.M. Hayden, G.G. Lonzarich and H.L. Skriver, Phys. Rev. B **33**, 4977 (1986).

[29] M.J. Steiner, F. Beckers, P.G. Nicklowitz and G.G. Lonzarich, Physica B **329**, 1079 (2003).

[30] J.J. Buiting, J. Klübler, and F.M. Mueller, J. Phys. F **39** L179 (1983).

[31] V.L. Moruzzi and P.M. Marcus, Phys. Rev. B, **42**, 5539 (1990).

[32] B.I. Min, A.J. Freeman, and H.J.F. Jansen, Phys. Rev. B **37**, 6757 (1988).

[33] J.H. Xu, B.I. Min, A.J. Freeman, and T. Oguchi, Phys. Rev. B **41**, 5010 (1990).

[34] G.Y Guo, Y.K. Wang, Li-Shing Hsu, J. Magn. Magn. Mater. **239**, 91 (2002).

[35] L.-S. Hsu, Y.-K. Wang and G.Y. Guo, J. Appl. Phys. **92**, 1419 (2002).

[36] D.J. Singh, *Planewaves Pseudopotentials and the LAPW Method* (Kluwer Academic, Boston, 1994).

[37] D. Singh, Phys. Rev. B **43**, 6388 (1991).

[38] S.H. Wei and H. Krakauer, Phys. Rev. Lett. **55**, 1200 (1985).

[39] P. Blaha, K. Schwarz G.K.H. Madsen, D. Kvasnicka, and J. Luitz, *WIEN2K, An Augmented Plane Wave + Local Orbitals Program for for Calculating Crystal Properties* (K. Schwarz, Techn. Universitat Wien, Austria, 2001), ISBN 3-9501031-1-2.

[40] L. Hedin and B. Lundqvist, J. Phys. C, **4**, 2064 (1971).

[41] U. von Barth and L. Hedin, J. Phys. C **5**, 1629 (1972).

[42] G.L. Krasko, Phys. Rev. B, **36** 8565 (1987).

[43] To compute pressure, we used $P = B/B'[(V/V_0)^{B'} - 1]$, where V/V_0 is the volume compression, B and B' are the bulk modulus and its derivative. We used the experimental bulk modulus of Ni_3Al, B=174 GPa. [46] For B' we used the calculated value B'=5.2.

[44] B. Efron and R.J. Tibshirani, *An Introduction to the Bootstrap* (Chapmann and Hall, New York, 1993).

[45] C.J. Schinkel, F.R. de Boer, and B. de Hon, J. Phys. F **3**, 1463 (1973).

[46] F. Wallow, G. Neite, W. Schroer and E. Nembach, Phys. Status Solidi A **99**, 483 (1987).

INTERPLAY BETWEEN HELICOIDAL MAGNETIC ORDERING AND SUPERCONDUCTIVITY ON THE DIFFERENTIAL CONDUCTANCE IN HONI$_2$B$_2$C/AG JUNCTIONS

I. N. Askerzade

Institute of Physics, Azerbaijan National Academy of Sciences, Baku-370143, Azerbaijan
and
Department of Physics, Ankara University, Tandogan, 06100, Ankara, Turkey

Abstract The point contact spectra of magnetic superconductor $HoNi_2B_2C$/Ag based junctions is analysed in the framework of Blonder-Tinkham-Klapwijk (BTK) theory. The anomalous behavior in the dI/dV curves above the Neel temperature ($T_N \sim 5\,\mathrm{K}$) is attempted to be explained by the partial suppression of superconducting gap parameter of the prevailing helical incommensurate structure.

Keywords: Magnetic superconductor, helical incommensurate structure

10.1 Introduction

Eight years after the discovery[1] of rare earth transition metal borocarbides (nitrides) RTBC(N) with T=Ni, Pd, Pt transition metals, the place of RTBC(N) compounds within the family of more or less exotic superconductors is still under debate. For this class of exotic superconductors there are several properties which taken together might be interpreted also as hints for unconventional (d-wave or p-wave) superconductivity. For example, d-wave superconductivity has been proposed for YNi_2B_2C and $LuNi_2B_2C$ compounds.[2] Phase-sensitive experiments [3] and the observation of Andreev bound state near appropriate surfaces [4] must await to confirm or disprove the predicted d-wave scenario.

It is well-known[5, 6] that the measurement of the differential conductivity of superconductor-insulator-normal metal (SC/I/N) junctions is a very sensitive method to probe the superconductinfg properties. Point-contact spectroscopy studies on borocarbide compounds are motivated by the possibility of a detailed investigation of the anisotropy of

S. Halilov (ed.), Physics of Spin in Solids: Materials, Methods and Applications, 155–160.

the gap parameter and the coexistence of superconductivity and magnetism in magnetic borocarbides. Andreev reflection spectroscopy for nonmagnetic borocarbides $Y(Lu)Ni_2B_2C$ is known to yield superconducting energy gap peaks. [7, 8] For the magnetic borocarbides with Dy and Er gap-like features in the Andreev reflection spectrum have also been seen. In the Dy compound superconductivity develops in the presence of antiferromagnetic ordering with $T_N = 10.5\,\mathrm{K}$ and it is the only borocarbide with the Neel temperature greater than the superconducting transition temperature, viz. $T_N > T_c = 6\,\mathrm{K}$. For instance, the Er compound with $T_c = 10.8\,\mathrm{K}$ exhibits antiferromagnetic (AFM) ordering [9] below $T_N = 5.9\,\mathrm{K}$. $HoNi_2B_2C$ compounds are marked by a complex magnetic structure.[10] In these compounds, the AFM structure develops below the Neel temperature $T_N \sim 5\,\mathrm{K}$ which is related[11] to the c-axis modulated commensurate magnetic structure with wave vector $Q_{AF} = c^* = 2\pi/c$. Other magnetic structures have been observed in the temperature region $T_N < T < T_m = 6\,\mathrm{K}$, spiral (helicoidal) c-axis modulated incommensurate with wave vector $Q_c = 0.91\,c^*$ and a-axis modulated incommensurate with wave vector $Q_a = 0.55\,a^*$. In $HoNi_2B_2C$ reentrant or almost reentrant superconductivity was detected over a large range manisfesting magnetic ordering.[12] Experimental point contact study was conducted by Rybaltchenko *et al.*[13] but the explanation of the suppression of Andreev pecularities are mostly unexplored.

Our primary aim is to discuss the influence of the helicoidal structyre on the $G_{NS}(V)$ curve of $HoNi_2B_2C/Ag$ junctions in the framework of Blonder, Tinkham, and Klapwijk[6] (BTK) formalism.

10.2 Basic Equations

First, we shall discuss the effect of a helicoidal structure on superconductivity. This question has been originally considered by Morosov [14] and also more recently in application to Ho borocarbides.[15] As it was shown[14, 15] using Bogoliubov transformations the gap parameter in the spectrum of electron quasiparticles becomes strongly anisotropic and vanishes at the boundaries of the breaks in the Fermi surface due to the Bragg planes generated by the magnetic ordering (i.e. when the Bragg planes intersect the Fermi surface).

Transport through NS junctions has successfully been investigated using the Bogoliubov-de Gennes (BdG) equation [6] . In the BdG formlism, the quasiparticles in SC are represented by a two-element column vector

$$\psi(\mathbf{x}) = \begin{pmatrix} u(\mathbf{x}) \\ v(\mathbf{x}) \end{pmatrix}, \tag{1}$$

where $u(\mathbf{x})$ and $v(\mathbf{x})$ are the electron and hole components of the quasi-particle excitations, and obey the BdG equations

$$Eu(\mathbf{x}) = H_0 u(\mathbf{x}) + \int d\mathbf{x}' \Delta(\mathbf{x}, \mathbf{x}') v(\mathbf{x}'), \tag{2}$$

$$Ev(\mathbf{x}) = -H_0 v(\mathbf{x}) + \int d\mathbf{x}' \Delta(\mathbf{x}, \mathbf{x}') u(\mathbf{x}'), \tag{3}$$

where $H_0 = -\frac{\nabla^2}{2m} + V(\mathbf{x}) - \mu$ is the single-particle Hamiltonian with μ being the Fermi energy , $V(\mathbf{x})$ and $\Delta(\mathbf{x}, \mathbf{x}')$ are the ordinary potential and pair potential, respectively. We assume the superconducting order parameter is not degraded by the normal metal, and thus neglect the proximity effect, i.e. for the NS interface (at x=0) problem we can write

$$\Delta(\mathbf{x}) = \Delta\Theta(x), \tag{4}$$

where $\Theta(x)$ is a step function. As result of calculations, the formula for the differential conductance of the junction normal metal-isotropic superconductor was obtained [6].

The BTK theory[6] for isotropic superconductors can be extended to the anisotropic case by including the momentum $\mathbf{k}$ depencence of the superconducting energy gap $\Delta(\mathbf{k})$ in the expression for Andreev reflection probabnility $A(\epsilon, \Delta(\mathbf{k}))$ and the normal reflection probability $B(\epsilon, \Delta(\mathbf{k}))$. Then, the differential conductance G_{NS} of an NS junction normalized to the normal state value G_{NN} at $T = 0$ can be written as

$$\frac{G_{NS}}{G_{NN}} = \frac{\frac{\partial I_{NS}}{\partial V}}{\frac{\partial I_{NN}}{\partial V}} = \frac{\frac{\partial}{\partial V} \int d^3\mathbf{k}\, v_z \left[1 + A(\epsilon, \Delta(\mathbf{k})) - B(\epsilon, \Delta(\mathbf{k}))\right]}{\frac{\partial}{\partial V} \int d^3\mathbf{k}\, v_z \left[1 - Z^2/(1 + Z^2)\right]}, \tag{5}$$

where Z is the barrier height, which can be introduced phenomenologically. v_z is the positive velocity component perpendicular to the interface of NS junction. As mentioned above, in this approximation the proximity effect is not taken into account, although the symmetry of the gap parameter strongly influences the behavior at the surface in the case of pure d-wave or p-wave symmetry.

Calculation of the differential conductance based on Eq. (1) for an N/d-wave superconductor has been performed by Tanaka *et al.*[16] and for a ferromagnet/d-wave superconductor by Zhu *et al.*[17] Dependence of the subgap structure on d-wave parameters and orientation of the to the NS boundary are presented. It is necessary to note calculations [18] in the framework of BTK formalism for possible p-wave gap parameter in Sr_2RuO_4. For the heavy fermion systems [19] UPt_3 similar calculations were performed by introducing odd-parity gap parameter. In all these cases the anisotropy of the gap parameter leads to a transformation of the plateu at $(-\Delta, \Delta)$ to triangular peak of the conductance in subgap region.

As pointed out by Morosov[15] the gap parameter of a superconductor in the presence of helical structure may be written as

$$\Delta(\mathbf{k}, T) = (u_{\mathbf{k}}^2 - v_{\mathbf{k}}^2)\Delta(T)\,, \tag{6}$$

where

$$u_{\mathbf{k}}^2 - v_{\mathbf{k}}^2 = \left(\frac{(\epsilon_{\mathbf{k}} - \epsilon_{\mathbf{k}+\mathbf{Q}})^2}{(\epsilon_{\mathbf{k}} - \epsilon_{\mathbf{k}+\mathbf{Q}})^2 + I^2 S^2} \right)^{1/2}\,, \tag{7}$$

in which I is the exchange interaction integral,S is the average ion spin,$\epsilon_{\mathbf{k}}$ is the dispersion telation in the partamagnetic phase and

$$\Delta(T) = \int_0^{\omega} d\epsilon\, \frac{\Delta(T)(1 - 2n_{\mathbf{k}})}{\epsilon^2 + \Delta^2(T)} \left(\int_{MFS} \frac{dS'}{(2\pi)^3} \frac{(u_{\mathbf{k}'}^2 - v_{\mathbf{k}'}^2)^2}{|\nabla_{\mathbf{k}'} \tilde{\epsilon}_{\mathbf{k}'}|} \right)\,, \tag{8}$$

where $\tilde{\epsilon}_{\mathbf{k}}$ is new dispersion relation[15] and $n_{\mathbf{k}}$ takes into account the accupation of the electronic state. The last equation corresponds to the usual BCS self-consistent gap equation with an effective parameter $\lambda_{\text{eff}}(T)$ defined as the term in brackets. $\lambda_{\text{eff}}(T)$ depends on the underlying magnetic state through the Bogoliubov coefficients and the slope of the magnetic Fermi surface.

10.3 Results and Discussions

Since all the anamolous magnetic wave vector dependencies come from the region where Fermi surface intersects the Bragg planes, the difference $\Delta\lambda(T) = \lambda - \lambda_{\text{eff}}(T)$ between the actual electron-phonon interaction λ and its effective value, we can expand quantities in terms of IS/ϵ_F. Using the results of band structure calculations[11] for borocarbide compounds the difference $\Delta\lambda(T)$ is estimated by Amici, Thalmeier, and Fulde[20] as $\Delta\lambda(T = 0)/\lambda = 0.12$. This result has been employed in the explanation of the main anomaly (reentrant behavior) of the upper critical field $H_{c_2}(T)$.

The experimental data for $HoNi_2B_2C$/Ag junctions show[13] insensitivity of the shape of $G_{NS}(V)$ curve to the orientation of the contact plane with respect to the crystal axis. This fact confirms the isotropic character of the electronic structure of these compounds. Thus, the possibility of d-wave or p-wave gap parameter in helical superconductors is elimintated. Evaluation of the shape of the $G_{NS}(V)$ curve for an NS junction by changing the parameter Z is analyzed by Blonder, Tinkham, and Klapwijk.[6] It is clear that the subgap plateau at $Z = 0$ transforms to two peaks at $\pm\Delta$ when the barrier height is increased. For the $HoNi_2B_2C/Ag$ contact at $T < 5\,K$, where helicoidal structure transforms into the antiferromagnetic phase, a double-peak structure is obtained [13].

However, in the temperature region $5 < T < 8,1\,K$, or equivalently $\Delta T/T \sim 3/8 \approx 0.4$, gapless behavior is observed (we remark that the corresponding value of the same parameter in $ErNi_2B_2C$ compound [9] is about ~ 0.2). In our opinion, the broadening character and the gapless behavior are related by the partial suppression of the order parameter in the presence of helical structure. As mentioned by Amici,Thalmeier, and Fulde[20] reduction of $\Delta\lambda/\lambda \sim 0,12$ is not sufficient for the total elimination of superconductivity and transition to the normal state. On the other hand, in calculating the $G_{NS}(V)$ curve using Eq. (1), we must take into account an additional reduction factor of $u_{\mathbf{k}}^2 - v_{\mathbf{k}}^2$. Because this latter factor is freater than unity, when averaged over the Fermi surface, we obtain an additional suppression of the gap parameter.

Thus, the broadening character of gapless behavior in $G_{NS}(V)$ curve at temperatures close to T_c for $HoNi_2B_2C/Ag$ junctions can be explained by the suppression of the gap parameter. Total suppression does not occur because the experiments of Rybaltchenko *et al.*[13] was conducted for a higly pure Ho compound. It follows from the experimental data [12] (resistive measurements) that in the region where helicoidal structure exists we have small (but not zero) gap parameter. As shown in [15] nonmagnetic impurities playes the important role in suppressing superconductivity in systems with helical magnetic structure. To put another way, due to the helical magnetic structure developing Andreev double-maximum structure is "delayed" in comparision with other magnetic borocarbides.

References

[1] Nagarajan R,Mazumdar C,Hossain Z,Dhar S.K,Gopalakrishnan K.V.,Gupta L.C.,Godart C.,Padalia B.D., and Vejayaraghavan R., 1994 Phys. Rev. Lett. **72** 274

[2] Wang G.,Maki K., 1998 Phys.Rev.B**58** 6493

[3] Wollman D.A.,Van Harlingen D.J.,Lee W.C.,Ginzberg D.M.,and Leggett A.J.,1993 Phys. Rev. Lett. **71** 2134

[4] Hu C.-R.,1994 Phys.Rev.Lett. **72** 1526

[5] Andreev A.F.,1964 Sov.Phys.-JETP**19** 1228

[6] Blonder G.E.,Tinkham M.,and Klapwjik T.M.1982 Phys.Rev.B **25** 4515

[7] Wilde Y.E., Iavarone M.,Welp W.,Metlushko V.,Koshelev A.E.,Aranson L.,and Grabtee G.W.,1997 Phys.Rev.Lett.**78** 4273

[8] Eskino T.,Fujii H.,Kosugi M.,Zeutrani Y.,and Akimitsu J., 1995 Phys.Rev.B **53** 5640

[9] Yanson I.K. 1999 in Symmetry and Pairing in superconductors (ed.by M.Ausloos and S.Kruchinin),Kluwer Academic Publisher,pp.271-285

[10] Goldman A.I.,Stassis S.,Canfield P.C.,Zarestky J., Dervenages P.,Cho B.K.,Johnston D.C., and Sternlieb B.,1994 Phys.Rev.B**50** 9668

[11] Drechsler S.L.,Shulga S.V.,Muller K.H.,Fuchs G.,Freudenberger J.,Behr G.,Eschrig H. ,Schultz L., Golden M.S., vonLips H., Fink J., Narozhnyi V.N. ,Rosner H., Zahn P.,Gladun A.,Lipp D., Kreyssig A., Loewenhaupt M., Koepernik K., Winzer K., Krug K., 1999 Physica C**317-318** 117

[12] Krug K.,Heinecke H.,and Winzer K. 1996 Physica C**26** **7**321

[13] Rybaltchenko L.F.,Jansen A.G.M.,Wyder P.,Tjutrina L.V.,Canfield P.C.,Tomy C.V., Pau D.McK., 1999 Physica C**319** 189

[14] Morosov A.I.,1980 Sov.Phys.Solid Stat.**22** 1974

[15] Morosov A.I., 1996Sov. Phys.- JETP **83**1048

[16] Tanaka Y.,and Kashiwaya S.,1995 Phys.Rev.Lett.**74** 3451

[17] Zhu J.-X., Freidman B.,and Ting C.S., 1999 Phys.Rev.B **59** 9558

[18] Honercamp C., and Sigrist M., 1998 J. Low Temp.Phys.**111** 895

[19] Goll G.,Bruder C., and Lohneysen H.v., 1995 Phys.Rev.B **52** 6801

[20] Amici A., Thalmeier P., and Fulde P., 2000 Phys. Rev. Lett. **84** 1800

AB INITIO CALCULATIONS OF THE OPTICAL AND MAGNETO-OPTICAL PROPERTIES OF MODERATELY CORRELATED SYSTEMS: ACCOUNTING FOR CORRELATION EFFECTS

A. Perlov, S. Chadov, H. Ebert
University of Munich, Butenandstrasse 5-13, D-81377, Munich, Germany

L. Chioncel, A. Lichtenstein
University of Nijmegen, NL-6526 ED Nijmegen, The Netherlands

M. Katsnelson
Uppsala University, P.O.Box 530, S-751 21 Uppsala, Sweden

Abstract The influence of dynamical correlation effects on the magneto-optical properties of ferromagnetic Fe and Ni has been investigated. In addition the temperature dependence of the self-energy and its influence on the DOS and optical conductivity is considered. Magneto-optical properties were calculated on the basis of the one-particle Green's function, which was obtained from the DMFT-SPTF procedure. It is shown that dynamical correlations play a rather important role in weakly correlated Fe and substantially change the spectra for moderately correlated Ni. Magneto-optical properties obtained for both systems are found in better agreement with experiment than by conventional LDA calculations.

Keywords: Dynamical correlations, DMFT, magneto-optics, self-energy.

11.1 Introduction

Much information on the electronic structure of magnetic solids is gained by optical and magneto-optical measurements, being useful tools for analyzing the dispersion of (quasi-particle) bands. However, measured optical and magneto-optical spectra can hardly be interpreted without accompanying theoretical calculations. For this purpose one in general has to solve a corresponding many-electron problem, which

S. Halilov (ed.), Physics of Spin in Solids: Materials, Methods and Applications, 161–176.

is impossible without the use of more or less severe approximations. For materials where the kinetic energy of the electrons is more important than the Coulomb interactions, the most successful first principles method is the Local (Spin-) Density Approximation (L(S)DA) to the Density Functional theory (DFT) [1], where the many-body problem is mapped onto a non-interacting system with a one-electron exchange-correlation potential approximated by that of the homogeneous electron gas. For the last two decades *ab initio* calculations of the optical and magneto-optical properties of solids based on this approximation yielded a good basis for such an interpretation, often leading to a quantitative agreement between theoretical and experimental spectra. The situation is very different when we consider more strongly correlated materials, (systems containing f and d electrons) since in all the calculations the LDA eigen-energies are implicitly interpreted to be the one-particle excitation energies of the system. It is well known that there are two possible sources of error connected with that approach: Firstly, the LDA provides only an approximate expression for the (local) exchange-correlation potential. Secondly, even with the exact exchange-correlation potential at hand, one is left with the problem that there is no known correspondence between the Kohn-Sham eigen-energies and the one-particle excitation energies [2–5].

For an in principle exact description of the excitation energies the non-local self-energy has to be considered. This, however, constitutes a many-body problem. Therefore, DFT-LDA calculations must be supplemented by many-body methods to arrive at a realistic description of the one-particle excitations in correlated systems. To give an example, let us mention the GW approximation [6] which is well suited for the case of insulators and semi-conductors and has also been applied successfully to transition metals [6–9]. Another approach is to consider Hubbard-type models where those Coulomb-interaction terms are included explicitly that are assumed to be treated insufficiently within DFT-LDA. Already the simplest Hartree-Fock like realization of such an approach called LDA+U [10] scheme allowed to improve considerably the description of the optical and magneto-optical spectra of strongly correlated systems (mostly containing rare earths elements [11, 12]). The main advantage of the LDA+U scheme is the energy independence of the self-energy which allows to use only slightly modified standard band structure methods for calculating optical and magneto-optical spectra. On the other hand the scheme works rather good only for extremely correlated systems, where Coulomb interactions (U) prevail considerably over the kinetic energy (bandwidth W). For moderately correlated systems (U≈W) which applies for most $3d$ and $5f$ elements and their compounds one has to take into account a non-Hermitian energy dependent self-energy to get a reasonable description of the electronic structure. Nowadays there are several approaches available to deal with this situation. The most advanced one is the Dynamical Mean-Field Theory (DMFT) [13]. DMFT is a successful approach to investigate strongly correlated systems with local Coulomb interactions. It uses the band structure results calculated, for example, within LDA approximation, as input and then missing electronic correlations are introduced by mapping the lattice problem onto an effective single-site problem which is equivalent to an Anderson impurity model [14]. Due to this equivalence a variety of approximative techniques have been used to solve the DMFT equations, such as Iterated Perturbation Theory (IPT) [13, 15], Non-Crossing Approximation (NCA) [16, 17], numerical techniques like Quantum Monte Carlo simulations (QMC) [18], Exact Diagonalization (ED) [15, 19], Numerical Renormalization Group (NRG) [20], or Fluctuation Exchange (FLEX) [21–23]. The DMFT maps lattice models onto quantum impurity models subject to a self-consistency condition in such a way that the many-body problem for the crystal splits into a single-particle impurity problem and a many-body problem of an effective atom. In fact, the DMFT, due to numerical and analytical techniques developed to solve the effective impurity problem [13], is a very efficient and extensively used approximation for energy-dependent self energy $\Sigma(\omega)$. At present LDA+DMFT is the only available *ab initio* computational technique which is able to treat correlated electronic systems close to a Mott-Hubbard MIT (Metal-Insulator Transition), heavy fermions and f-electron systems.

Concerning the calculation of the optical spectra we have to face the following problem: one particle wave functions are not defined any more and the formalism has to applied in the Green function representation. Such a representation has already been derived [24] and successfully applied for calculations in the framework of Korringa-Kohn-Rostoker (KKR) Green-function method for LSDA calculations. The only drawback of such an approach is that it is highly demanding as to both computational resources and computational time.

In this paper we propose a simplified way to calculate optical and magneto-optical properties of solids in the Green function representation based on variational methods of band structure calculations.

The paper is organized as following: in section 2 the formalism for Green's function calculations of optical and magneto-optical properties that account for many-body effects through an effective self-energy is presented. Then, the DMFT-SPTF method for the calculation of the self-energy is considered. In section 3 the obtained results of our calculations for Fe and Ni are discussed and compared with experimental ones. The last section 4 contains the conclusion and an outlook.

11.2 Green's function calculations of the conductivity tensor

Optical properties of solids are conventionally described in terms of either the dielectric function or the optical conductivity tensor which are connected via the simple relationship:

$$\sigma_{\alpha\beta}(\omega) = -\frac{i\omega}{4\pi}(\varepsilon_{\alpha\beta}(\omega) - \delta_{\alpha\beta})\,. \tag{1}$$

The optical conductivity is connected directly to the other optical properties. For example, the Kerr rotation $\theta_K(\omega)$ and so-called Kerr ellipticity $\varepsilon_K(\omega)$ for small angles and $\mid \varepsilon_{xy} \mid \ll \mid \varepsilon_{xx} \mid$ can be calculated using the expression [25]:

$$\theta_K(\omega) + i\varepsilon_K(\omega) = \frac{-\,\sigma_{xy}(\omega)}{\sigma_{xx}(\omega)\left[1 + \frac{4\pi}{\omega}\sigma_{xx}(\omega)\right]^{1/2}}\;. \tag{2}$$

The reflectivity coefficient r is given by

$$r = \frac{(n-1)^2 + k^2}{(n+1)^2 + k^2} \tag{3}$$

with n and k being the components of the complex refractive index, namely refractive and absorptive indices, respectively. They are connected to the dielectric function via:

$$n + ik = (\varepsilon_{xx} + i\varepsilon_{xy})^{1/2}\;. \tag{4}$$

Microscopic calculations of the optical conductivity tensor are based on the Kubo linear response formalism [26]:

$$\sigma_{\alpha\beta}(\omega) = -\frac{1}{\hbar\omega V}\int_{-\infty}^{0} d\tau e^{-i(\omega+i\eta)\tau} \langle\ [J_\beta(\tau), J_\alpha(0)]\ \rangle \tag{5}$$

involving the expectation value of the correlator of the electric current operator $J_\alpha(\tau)$. In the framework of the quasiparticle description of the excitation spectra of solids the formula can be rewritten in the spirit of the Greenwood approach and making use of the one-particle Green function $G(E)$:

$$\sigma_{\alpha\beta}(\omega) = \frac{i\hbar}{\pi^2 V}\int_{-\infty}^{\infty} dE \int_{-\infty}^{\infty} dE' f(E-\mu) f(\mu - E')$$

$$\left[\frac{\mathrm{Tr}\left\{\hat{j}_\alpha \Im G(E')\hat{j}_\beta \Im G(E)\right\}}{(E'-E+i\eta)(\hbar\omega+E-E'+i\eta)} + \frac{\mathrm{Tr}\left\{\hat{j}_\beta \Im G(E')\hat{j}_\alpha \Im G(E)\right\}}{(E'-E+i\eta)(\hbar\omega+E'-E+i\eta)}\right], \tag{6}$$

where $\Im G(E)$ stands for the anti-Hermitian part of the Green's function, $f(E)$ is the Fermi function and V is the volume of a sample. Taking the zero temperature limit and making use of the analytical properties of the Green's function one can get a simpler expression for the absorptive (anti-Hermitian) part of the conductivity tensor:

$$\sigma^{(1)}_{\alpha\beta}(\omega) = \frac{1}{\pi\omega}\int_{E_F-\omega}^{E_F} dE\ \mathrm{tr}\left[\hat{j}_\alpha \Im G(E)\hat{j}_\beta \Im G(E+\hbar\omega)\right]. \tag{7}$$

The dispersive part of $\sigma_{\alpha\beta}(\omega)$ is connected to the absorptive one via a Kramers-Kronig relationship.

The central quantity entering expression Eq.(7) is the one-particle Green's function defined as a solution of the equation:

$$[\hat{H}_0 + \hat{\Sigma}(E) - E]\hat{G}(E) = \hat{I}\,, \tag{8}$$

where $\hat{H}_0$ is a one-particle Hamiltonian including the kinetic energy, the electron-ion Coulomb interaction and the Hartree potential, while the self-energy $\hat{\Sigma}(E)$ describes all static and dynamic effects of electron-electron exchange and correlations. The L(S)DA introduces the self-energy as a local, energy independent exchange-correlation potential $V_{xc}(r)$. As the introduction of such an additional potential does not change the properties of $\hat{H}_0$ we will incorporate this potential to $\hat{H}_{LDA}$ and subtract this term from the self-energy operator. This means that the self energy Σ used in the following is meant to describe exchange and correlation effects not accounted for within LSDA.

With a choice of the complete basis set $\{|i\rangle\}$ the Green's function can be represented as:

$$G(E) = \sum_{ij} |i\rangle G_{ij}(E)\langle j|\,, \tag{9}$$

with the Green's matrix G_{ij} being defined as

$$G_{ij}(E) = \left[\langle i|\hat{H}|j\rangle - E\langle i|j\rangle - \langle i|\hat{\Sigma}(E)|j\rangle\right]^{-1}. \tag{10}$$

Dealing with crystals one can make use of Bloch's theorem when choosing basic functions $|i_{\mathbf{k}}\rangle$. This leads to the $\mathbf{k}$-dependent Green's function matrix

$$G_{ij}^{\mathbf{k}}(E) = [H_{ij}^{\mathbf{k}} - EO_{ij}^{\mathbf{k}} - \Sigma_{ij}^{\mathbf{k}}(E)]^{-1} \,. \tag{11}$$

Introducing the anti-Hermitian part of the Green's function matrix as

$$\mathcal{G}_{ij}^{\mathbf{k}}(E) = \frac{i}{2}[G_{ij}^{\mathbf{k}}(E) - G_{ji}^{\mathbf{k}}(E)] \tag{12}$$

and taking into account the above mentioned translational symmetry we obtain the following expression for the absorptive part of the optical conductivity:

$$\sigma_{\alpha\beta}^{abs} = \frac{1}{\pi\omega}\int_{E_F-\hbar\omega}^{E_F} dE \int d^3k \sum_{ij} \mathcal{J}_{ij}^{\alpha}(\mathbf{k},E)\mathcal{J}_{ji}^{\beta}(\mathbf{k},E+\hbar\omega) \tag{13}$$

with

$$\mathcal{J}_{ij}^{\alpha}(\mathbf{k},E) = \sum_{n} \mathcal{G}_{in}^{\mathbf{k}}(E)\langle n_{\mathbf{k}}|\hat{j}^{\alpha}|j_{\mathbf{k}}\rangle \tag{14}$$

The efficiency and accuracy of the approach is determined by the choice of $|i_{\mathbf{k}}\rangle$. One of the computationally most efficient variational methods is the Linear Muffin-Tin Orbitals method [27] which allows one to get a rather accurate description of the valence/conduction band in the range of about 1 Ry, which is enough for the calculations of the optical spectra ($\hbar\omega < 6 - 8$ eV). This method has been used in the present work. A detailed description of the application of the above sketched approach in the framework of LMTO can be found elsewhere [28].

Calculation of the self-energy

The key point for accounting of many-body correlations in the present approach is the choice of approximation for the self-energy. As it was discussed in the Introduction one of the most elaborated modern approximation is DMFT.

For the present work we have chosen one of the most computationally efficient variants of DMFT: Spin polarized T-matrix plus fluctuation exchange (SPTF) approximation [23], which is based on the general many-body Hamiltonian in the LDA+U scheme:

$$\begin{aligned} H &= H_t + H_U \\ H_t &= \sum_{\lambda\lambda'\sigma} t_{\lambda\lambda'} c_{\lambda\sigma}^{+} c_{\lambda'\sigma} \\ H_U &= \frac{1}{2}\sum_{\{\lambda_i\}\sigma\sigma'} \langle\lambda_1\lambda_2\,|v|\,\lambda_1'\lambda_2'\rangle\, c_{\lambda_1\sigma}^{+} c_{\lambda_2\sigma'}^{+} c_{\lambda_2'\sigma'} c_{\lambda_1'\sigma}\,, \end{aligned} \tag{15}$$

where $\lambda = im$ are the site number (i) and orbital (m) quantum numbers, $\sigma =\uparrow,\downarrow$ is the spin projection, c^+, c are the Fermion creation and annihilation operators, H_t is the effective single-particle Hamiltonian from the LDA, corrected for the double-counting of average interactions among correlated electrons as it will be described below. The matrix elements of the screened Coulomb potential are defined in the standard way

$$\langle 12\,|v|\,34\rangle = \int d\mathbf{r}d\mathbf{r}'\psi_1^*(\mathbf{r})\psi_2^*(\mathbf{r}')v\left(\mathbf{r}-\mathbf{r}'\right)\psi_3(\mathbf{r})\psi_4(\mathbf{r}'), \tag{16}$$

where we define for briefness $\lambda_1 \equiv 1$ etc. A general SPTF scheme has been presented recently [23]. For d electrons in cubic structures where the one-site Green function is diagonal in orbital indices the general formalism can be simplified. First, the basic equation for the T-matrix which replaces the effective potential in the SPTF approach reads

$$\begin{aligned}\left\langle 13\left|T^{\sigma\sigma'}(i\Omega)\right|24\right\rangle = \langle 13\,|v|\,24\rangle \frac{1}{\beta}\sum_{\omega}\sum_{56}\langle 13\,|v|\,56\rangle \times \\ G_5^{\sigma}(i\omega)\,G_6^{\sigma'}(i\Omega - i\omega)\left\langle 56\left|T^{\sigma\sigma'}(i\Omega)\right|24\right\rangle,\end{aligned} \tag{17}$$

where $\omega = (2n+1)\pi T$ are the Matsubara frequencies for temperature $T \equiv \beta^{-1} \quad (n = 0, \pm 1, ...)$.

At first, we should take into account the "Hartree" and "Fock" diagrams with the replacement of the bare interaction by the T-matrix

$$\begin{aligned}\Sigma_{12,\sigma}^{(TH)}(i\omega) &= \frac{1}{\beta}\sum_{\Omega}\sum_{3\sigma'}\left\langle 13\left|T^{\sigma\sigma'}(i\Omega)\right|23\right\rangle G_3^{\sigma'}(i\Omega - i\omega) \\ \Sigma_{12,\sigma}^{(TF)}(i\omega) &= -\frac{1}{\beta}\sum_{\Omega}\sum_{3}\langle 13\,|T^{\sigma\sigma}(i\Omega)|\,32\rangle\, G_3^{\sigma}(i\Omega - i\omega)\,.\end{aligned} \tag{18}$$

Now we rewrite the effective Hamiltonian (15) with the replacement $\langle 12\,|v|\,34\rangle$ by $\left\langle 12\left|T^{\sigma\sigma'}\right|34\right\rangle$ in H_U. To consider the correlation effects described due to P-H channel we have to separate density (d) and magnetic (m) channels as in Ref.[21]

$$\begin{aligned}d_{12} &= \frac{1}{\sqrt{2}}\left(c_{1\uparrow}^{+}c_{2\uparrow} + c_{1\downarrow}^{+}c_{2\downarrow}\right) \\ m_{12}^{0} &= \frac{1}{\sqrt{2}}\left(c_{1\uparrow}^{+}c_{2\uparrow} - c_{1\downarrow}^{+}c_{2\downarrow}\right) \\ m_{12}^{+} &= c_{1\uparrow}^{+}c_{2\downarrow} \\ m_{12}^{-} &= c_{1\downarrow}^{+}c_{2\uparrow}\,.\end{aligned} \tag{19}$$

Then the interaction Hamiltonian can be rewritten in the following matrix form

$$H_U = \frac{1}{2} Tr\left(D^+ * V^{\|} * D + m^+ * V_m^{\perp} * m^- + m^- * V_m^{\perp} * m^+ \right), \quad (20)$$

where $*$ means the matrix multiplication with respect to the pairs of orbital indices, e.g.

$$\left(V_m^{\perp} * m^+ \right)_{11'} = \sum_{34} \left(V_m^{\perp} \right)_{11',22'} m_{22'}^+ \, .$$

The supervector D is defined as

$$D = \left(d, m^0 \right), D^+ = \left(\begin{array}{c} d^+ \\ m_0^+ \end{array} \right),$$

and the effective interactions have the following form:

$$\left(V_m^{\perp} \right)_{11',22'} = -\left\langle 12 \left| T^{\uparrow\downarrow} \right| 2'1' \right\rangle$$

$$V^{\|} = \left(\begin{array}{cc} V^{dd} & V^{dm} \\ V^{md} & V^{dd} \end{array} \right)$$

$$V_{11',22'}^{dd} = \frac{1}{2} \sum_{\sigma\sigma'} \left\langle 12 \left| T^{\sigma\sigma'} \right| 1'2' \right\rangle - \frac{1}{2} \sum_{\sigma} \left\langle 12 \left| T^{\sigma\sigma} \right| 2'1' \right\rangle$$

$$V_{11',22'}^{mm} = \frac{1}{2} \sum_{\sigma\sigma'} \sigma\sigma' \left\langle 12 \left| T^{\sigma\sigma'} \right| 1'2' \right\rangle - \frac{1}{2} \sum_{\sigma} \left\langle 12 \left| T^{\sigma\sigma} \right| 2'1' \right\rangle$$

$$V_{11',22'}^{dm} = V_{22',11'}^{md} =$$
$$\frac{1}{2} [\left\langle 12 \left| T^{\uparrow\uparrow} \right| 1'2' \right\rangle - \left\langle 12 \left| T^{\downarrow\downarrow} \right| 1'2' \right\rangle - \left\langle 12 \left| T^{\uparrow\downarrow} \right| 1'2' \right\rangle$$
$$+ \left\langle 12 \left| T^{\downarrow\uparrow} \right| 1'2' \right\rangle - \left\langle 12 \left| T^{\uparrow\uparrow} \right| 2'1' \right\rangle + \left\langle 12 \left| T^{\downarrow\downarrow} \right| 2'1' \right\rangle] \, . \quad (21)$$

To calculate the particle-hole (P-H) contribution to the electron self-energy we first have to write the expressions for the generalized susceptibilities, both transverse $\chi^{\perp}$ and longitudinal $\chi^{\|}$. One has

$$\chi^{+-}(i\omega) = \left[1 + V_m^{\perp} * \Gamma^{\uparrow\downarrow}(i\omega) \right]^{-1} * \Gamma^{\uparrow\downarrow}(i\omega) \, , \quad (22)$$

where

$$\Gamma_{12,34}^{\sigma\sigma'}(\tau) = -G_2^{\sigma}(\tau) \, G_1^{\sigma'}(-\tau) \, \delta_{23}\delta_{14} \quad (23)$$

is an "empty loop" susceptibility and $\Gamma(i\omega)$ is its Fourier transform, τ is the imaginary time. The corresponding longitudinal susceptibility matrix has a more complicated form:

$$\chi^{\|}(i\omega) = \left[1 + V^{\|} * \chi_0^{\|}(i\omega) \right]^{-1} * \chi_0^{\|}(i\omega), \quad (24)$$

and the matrix of the bare longitudinal susceptibility is

$$\chi_0^{\parallel} = \frac{1}{2}\begin{pmatrix} \Gamma^{\uparrow\uparrow} + \Gamma^{\downarrow\downarrow} & \Gamma^{\uparrow\uparrow} - \Gamma^{\downarrow\downarrow} \\ \Gamma^{\uparrow\uparrow} - \Gamma^{\downarrow\downarrow} & \Gamma^{\uparrow\uparrow} + \Gamma^{\downarrow\downarrow} \end{pmatrix}, \tag{25}$$

in the *dd*-, dm^0-, m^0d-, and m^0m^0- channels ($d, m^0 = 1, 2$ in the supermatrix indices). An important feature of these equations is the coupling of longitudinal magnetic fluctuations and of density fluctuations. It is not present in the one-band Hubbard model due to the absence of the interaction of electrons with parallel spins. For this case Eqs. (22) and (24) coincide with the well-known result of Izuyama *et. al.* [29].

Now we can write the particle-hole contribution to the self-energy. Similar to Ref.[22] one has

$$\Sigma_{12,\sigma}^{(ph)}(\tau) = \sum_{34,\sigma'} W_{13,42}^{\sigma\sigma'}(\tau)\, G_{34}^{\sigma'}(\tau), \tag{26}$$

with the P-H fluctuation potential matrix:

$$W^{\sigma\sigma'}(i\omega) = \begin{bmatrix} W^{\uparrow\uparrow}(i\omega) & W^{\perp}(i\omega) \\ W^{\perp}(i\omega) & W^{\downarrow\downarrow}(i\omega) \end{bmatrix}, \tag{27}$$

were the spin-dependent effective potentials are defined as

$$\begin{aligned} W^{\uparrow\uparrow} &= \frac{1}{2}V^{\parallel} * \left[\chi^{\parallel} - \chi_0^{\parallel}\right] * V^{\parallel} \\ W^{\downarrow\downarrow} &= \frac{1}{2}V^{\parallel} * \left[\widetilde{\chi}^{\parallel} - \widetilde{\chi}_0^{\parallel}\right] * V^{\parallel} \\ W^{\uparrow\downarrow} &= V_m^{\perp} * \left[\chi^{+-} - \chi_0^{+-}\right] * V_m^{\perp} \\ W^{\downarrow\uparrow} &= V_m^{\perp} * \left[\chi^{-+} - \chi_0^{-+}\right] * V_m^{\perp}\,. \end{aligned} \tag{28}$$

Here $\widetilde{\chi}^{\parallel}, \widetilde{\chi}_0^{\parallel}$ differ from $\chi^{\parallel}, \chi_0^{\parallel}$ by the replacement of $\Gamma^{\uparrow\uparrow} \Leftrightarrow \Gamma^{\downarrow\downarrow}$ in Eq.(25). We have subtracted the second-order contributions since they have already been taken into account in Eq.(18).

Our final expression for the self energy is

$$\Sigma = \Sigma^{(TH)} + \Sigma^{(TF)} + \Sigma^{(PH)}\,. \tag{29}$$

This formulation takes into account accurately spin-polaron effects because of the interaction with magnetic fluctuations [30, 31], the energy dependence of the T-matrix which is important for describing the satellite effects in Ni [32], contains exact second-order terms in v and is rigorous (because of the first term) for almost filled or almost empty bands.

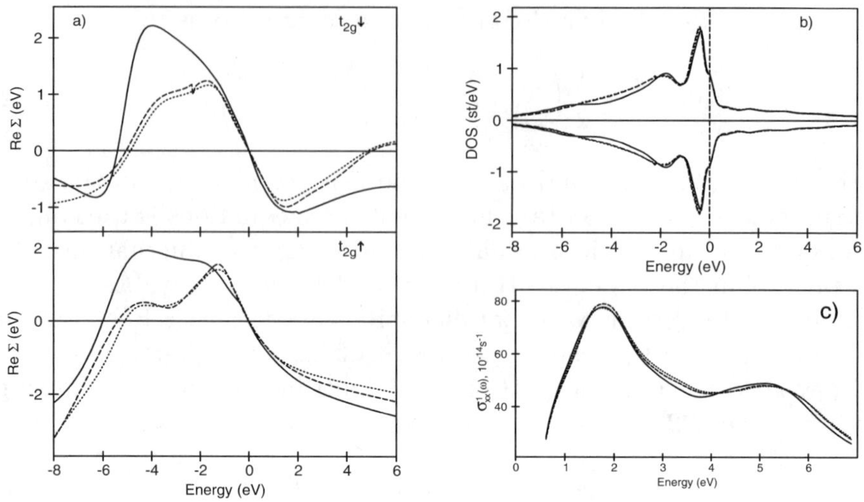

Figure 1. The self-energy (a) of Fe for three different temperatures and corresponding densities of states (b) and optical conductivities spectra (c). Full, dashed and dotted lines correspond to $T = 125K$, $T = 300K$ and $T = 900K$, respectively.

Since the LSDA Green's function already contains the average electron-electron interaction, in Eqs. (18) and (26) the static part of the self-energy $\Sigma^\sigma(0)$ is not included, i.e. we have

$$\tilde{\Sigma}^\sigma(i\omega) = \Sigma^\sigma(i\omega) - \Sigma^\sigma(0). \tag{30}$$

11.3 Results and discussion

The matrix elements of v appearing in Eq.(16) can be calculated in terms of two parameters - the averaged screened Coulomb interaction U and exchange interaction J [23]. The screening of the exchange interaction is usually small and the value of J can be calculated directly. Moreover numeric calculations show that the value of J for all $3d$ elements is practically the same and approximately equal to 0.9 eV. This value has been adopted for all our calculations presented here. At the same time direct Coulomb interaction undergoes substantional screening and one has to be extremely careful making the choice for this parameter. There are some prescription how one can get it within constraint LDA calculation [2]. However, results obtained in this way depend noticeably on the choice of the basis functions, way of accounting for hybridization etc. Nevertheless the order of magnitude coming out from various ap-

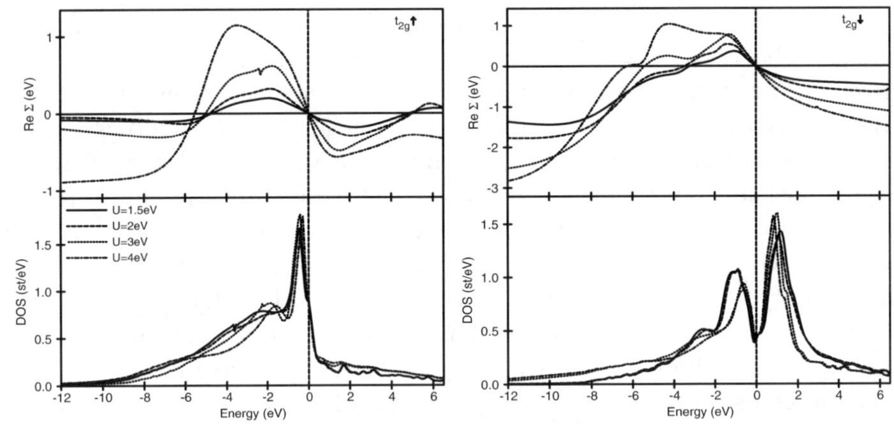

Figure 2. The real part of t_{2g} component of Σ for $T = 300K$ in Fe for various values of U; left: spin-up, right: spin-down.

proaches is the same giving the value of U in the range 1–4 eV. In the present paper we are discussing the influence of the choice of U on the calculated optical spectra.

Another parameter entering SPTF equations is temperature. For a moment we are more interested in the low temperature properties while computationally the higher the tempreture the less computationally demanding are the calculations. This is why we decided first to consider the dependence of the self-energy on the temperature.

In Fig. 1 we show the self-energy obtained for Fe for three different temperatures as well as corresponding densities of states and optical conductivities spectra. One can see that despite the differences in Σ are quite noticeable this leads only to moderate changes in the density of states and does not affect the optical conductivity.

Much more important for the results is the parameter U. Fig. 2 shows as an example the real part t_{2g} component of Σ for $T = 300K$ in Fe for various values of U. Despite the overall shape of the curve is practically the same the magnitude of the self-energy increasing with increase of U as it is expected from the analytical expressions. This change in self-energy leads to corresponding changes in the densities of states especially noticeable for the minority spin subband. The influence of the choice of U on the optical properties is even more pronounced (see Fig. 3). The low energy peak in the diagonal part of the optical conductivity shifts to the lower energies reaching the experimental position already for U=1.5 eV. In the high energy part of the spectra large values of U lead to a structure around 5 eV not seen in experiment.

Again, the value U=1.5 eV gives also the best description for the shape of the experimental curve. (Note, that the experimental results for $\sigma^1_{xx}(\omega)$ are multiplied by a factor of 1.7 to make the comparison more obvious.) A rather different situation occurs for the off-diagonal part of the optical conductivity. The low energy peak can be brought to the proper position only with U=4 eV, at the same time the shape of the theoretical curves above 2.5 eV has a rather different structure in comparison with the experimental one, only crossing the zero axis at the same energy. However, a direct comparison of calculated σ_{xy} data with experimental ones may be somewhat misleading as experimentally this quantity cannot be measured directly and is usually obtained from ellipsometric measurements and measurements of the Kerr rotation spectra. Thus in the left panel of the Fig. 3 we show our results for the calculated polar Kerr rotation spectra in comparison with experimental data. As one can see again the DMFT calculation with U = 1.5 eV describes the experimental data in a rather satisfactory way.

If for Fe LSDA calculations already give a reasonable description of the optical properties and the many-body correlation effects, which improves only minor details, the situation in Ni is quite different. It is well-known that LSDA fails to describe the bandwidth for Ni, causing problems in the theoretical interpretation of all the spectroscopic experiments such as photoemission, x-ray emission, optics, etc. The main reason for this is the underestimation of electron-electron correlations which appear to be relatively strong in this metal. Again, as in the case of Fe, we carried out calculations with different values of U to find the best description of

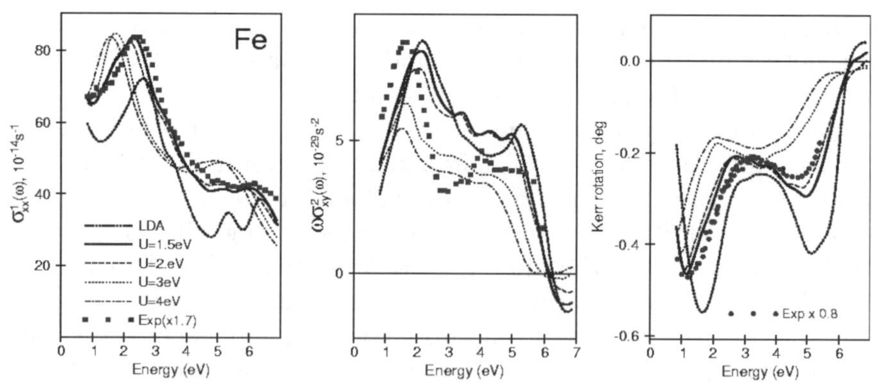

Figure 3. Optical conductivity (left: diagonal; middle: off-diagonal) and polar Kerr rotation (right) spectra in comparison with the experimental data of Fe. Experimental data for conductivity are taken from Ref. [33, 34]; Kerr rotation spectra - from Ref. [35]

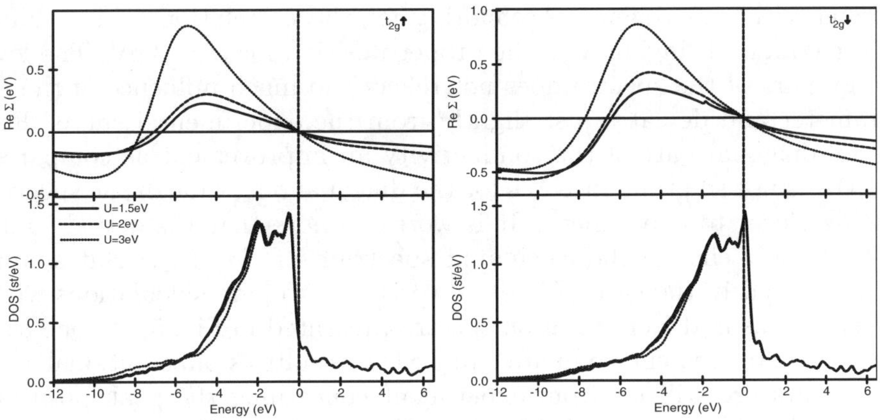

Figure 4. The real part of t_{2g} self-energy for $U = 1.5, 2, 3$ eV and corresponding DOS plots for Ni; left: spin-up, right: spin-down.

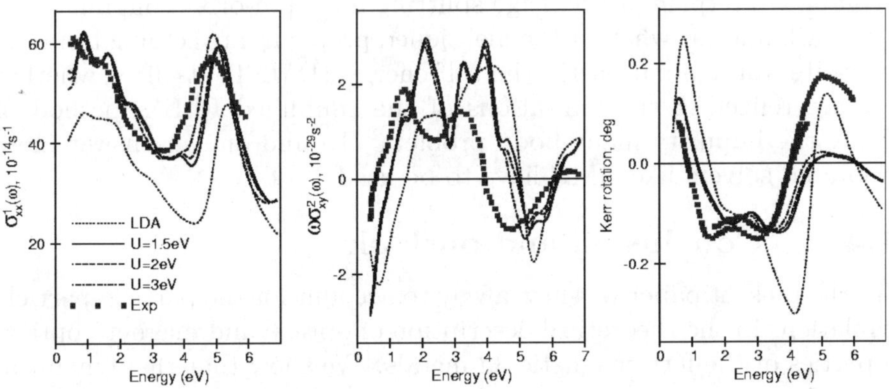

Figure 5. Optical conductivity and polar Kerr rotation spectra in comparison with the experimental data of Ni. Experimental data for conductivity are taken from Ref. [36, 37]. Kerr rotation spectra - from Ref. [35]

the spectral properties of Ni. In Fig. 4 we show the real part of the t_{2g} self-energy for $U = 1.5, 2, 3$ eV as well as corresponding DOS plots.

Despite the changes in the amplitude of the self-energy are huge, all self-energies lead to rather small changes in the density of states, narrowing somewhat the bandwidth only and developing a low energy tail. Nevertheless the diagonal part of the conductivity which reflects the convolution of the occupied and unoccupied states is much more affected by the choice of U. The main change can be seen in the position of high energy peak which is placed by LDA about 1 eV higher in comparison with experiment. Accounting for the correlation effects shifts this maximum bringing it to the proper position for $U = 3$ eV. The low energy part of the spectra does not reflect too much influence of the U parameter and deviates just slightly from the experimental curve. For the off-diagonal part of the conductivity an improvement as compared to LDA is not so pronounced as for the diagonal one, though the spectra getting closer to experiment. It is worth to note that the actual value of U doesn't change the calculated spectrum of $\omega\sigma_{xy}^2(\omega)$. But again, as mentioned in the case of Fe it is worth to compare calculations with directly measured Kerr rotation spectra presented in Fig. 5. As one can see, the improvement compared to LSDA results is substantional but our results are still far from experiment concerning the peak position both in the infrared and visible parts of the spectra. This disagreement is apparently coming from the approximation that has been made and is much more pronounced in the off-diagonal part of conductivity as it is more sensitive to the details of the electronic structure being the result of complex interplay of exchange splitting and spin-orbit coupling.

It is still unclear whether the mentioned problems are coming from the single-site approximation for the self-energy (DMFT) itself or whether they are reflecting the limitations of the simplified FLEX method of solving the impurity many-body problem. To find out an answer more elaborated solvers like QMC have to be used.

11.4 Conclusion and outlook

In the present paper we show a way to account for the particle-particle correlations in the theoretical description of optical and magneto-optical properties of the ferromagnetic 3*d* metals. We show that the dynamical correlations play an important role even in weakly correlated materials like Fe and can substantially change the shape of the spectra for moderately correlated Ni. Even a rather simple way of accounting for dynamic correlation allows to improve theoretical results substantionally though not giving the perfect agreement with experiment.

Thus to go further one has to use more elaborated technique to obtain the self-energy both within DMFT and beyond (for example, new DMFT+GW approximation). Work along this line is in progress.

References

[1] P. Hohenberg and W. Kohn, Phys. Rev., **136**, B864 (1964).

[2] R. O. Jones and O. Gunnarsson, Rev. Mod. Phys. **61**, 689 (1989).

[3] W. Kohn and P. Vashishta, in: *Theory of the Inhomogeneous Electron Gas*, Ed. by S. Lundqvist and N. H. March, p. 79. Plenum, New York (1983).

[4] C.-O. Almbladh and U. von Barth, in: *Density Functional Methods in Physics*, Ed. by R. M. Dreizler and J. da Providencia, p. 209. Plenum, New York (1985).

[5] G. Borstel, Appl. Phys. A **38**, 193 (1985).

[6] L. Hedin, Phys. Rev. **139**, 796 (1965).

[7] F. Aryasetiawan, Phys. Rev. B **46**, 13051 (1992).

[8] M. Rohlfing, P. Krüger, and J. Pollmann, Phys. Rev. B **48**, 17791 (1993).

[9] F. Aryasetiawan and O. Gunnarsson, Phys. Rev. Lett. **74**, 3221 (1995).

[10] Anisimov V. I., Zaanen J. and Andersen O. K., Phys. Rev. B, **44** (1991) 943.

[11] A.N. Yaresko, P.M. Oppeneer, A.Ya. Perlov, V.N. Antonov, T. Kraft, and H. Eschrig, Europhys. Lett. **36**, 551, (1996).

[12] V. N. Antonov, V. P. Antropov, B. N. Harmon, A. N. Yaresko, and A. Ya. Perlov Phys. Rev. B **59**, 14552,(1999).

[13] A. Georges, G. Kotliar, W. Krauth, and M. Rozenberg, Rev. Mod. Phys. **68**, 13 (1996).

[14] P. W. Anderson, in *Moment formation in solids*, edited by W. J. L. Buyers, Plenum Press, New York and London, 1984, p. 313.

[15] A. Georges, G. Kotliar, Phys. Rev. B **45**, 6479 (1992).

[16] H. Keiter, J. C. Kimball, Phys. Rev. Lett., **25**, 672 (1970)

[17] N. E. Bickers, D. L. Cox, J. W. Wilkins, Phys. Rev. B **36**, 2036 (1987)

[18] J. E. Hirsch, R. M. Fye, Phys. Rev. Lett. **56**, 2521 (1986); M. Jarrel, Phys. Rev. Lett. **69**, 168 (1992); M. Rozenberg, X. Y. Zhang, G. Kotliar, Phys. Rev. Lett. **69**, 1236 (1992); A. Georges, W. Krauth, Phys. Rev. Lett. **69**, 1240 (1992).

[19] M. Caffarel, W. Krauth, Phys. Rev. Lett. **72**, 1545 (1994)

[20] R. Bulla, Adv. Sol. State Phys. **46**, 169 (2000)

[21] N. E. Bickers, D. J. Scalapino, Ann. Phys. (N.Y.) **193**, 206 (1989)

[22] M. I. Katsnelson and A. I. Lichtenstein, J. Phys.: Condens. Matter **11**, 1037 (1999)

[23] M. I. Katsnelson and A. I. Lichtenstein, Eur. Phys. J. B **30**, 9 (2002)

[24] T. Huhne, H. Ebert, Phys. Rev. B **60**, 12982 (1999).

[25] V. N. Antonov, A. N. Yaresko, A. Ya. Perlov, V. V. Nemoshkalenko, Low Temperature Physics **25**, 387 (1999)

[26] R. Kubo, J. Phys. Soc. Jpn. **71**, 585 (1958)

[27] O. K. Andersen, Phys. Rev. B **12**, 3060 (1975).

[28] A. Perlov, S. Chadov, H. Ebert, Phys. Rev. B **68**, 245112 (2003).

[29] T. Izuyama, D. Kim, and R. Kubo, J. Phys. Soc. Japan 18, 1025 (1963).

[30] Yu. Irkhin and M. I. Katsnelson, J. Phys. : Condens. Matter 2, 7151 (1990). 37. H. A. Mook and D. McK. Paul, Phys. Rev. Lett.54, 227

[31] V. Yu. Irkhin and M. I. Katsnelson, Physics - Uspekhi **37**,659 (1994).

[32] A. Liebsch, Phys. Rev. B 23 , 5203 (1981).

[33] H.T. Yolken, J. Kruger, J. Opt. Soc. Am. **55**, 842 (1965)

[34] G. S. Krichnik, V. A. Artem'ev, Sov. Phys. J.E.T.P. **26**, 1080 (1968)

[35] P.G. van Engen, Ph.D. thesis, Tech. Univ. of Delft (1983), S. Visnovsky *et al.*, J. Magn. Magn. Materials **127**, 135 (1993)

[36] P. B. Johnson, R. W. Christy, Phys. Rev. B **9**, 5056 (1974)

[37] J. L. Erskine, Physica B **89**, 83 (1977)

SPIN-DEPENDENT TRANSPORT IN PHASE-SEPARATED MANGANITES

K. I. Kugel, A. L. Rakhmanov, A. O. Sboychakov
Institute for Theoretical and Applied Electrodynamics, Russian Academy of Sciences
Izhorskaya str. 13/19, Moscow 125412, Russia
kugel@orc.ru

M. Yu. Kagan, I. V. Brodsky, A. V. Klaptsov
Kapitza Institute for Physical Problems, Russian Academy of Sciences
Kosygina str. 2, Moscow, 119334 Russia

Abstract Starting from the assumption that ferromagnetically correlated regions exist in manganites even in the absence of long-range magnetic order, we construct a model of charge transfer due to the spin-dependent tunnelling of charge carriers between such regions. This model allows us to analyze the temperature and magnetic field dependence of resistivity, magnetoresistance, and magnetic susceptibility of phase-separated manganites in the temperature range corresponding to non-metallic behavior. The comparison of theoretical and experimental results reveals the main characteristics of the phase-separated state.

Keywords: manganites, phase separation, spin-dependent tunnelling

12.1 Introduction

Unusual properties and the richness of the phase diagram of manganites gave rise to a huge number of papers dealing with different aspects of the physics of these compounds. A special current interest to manganites is related to the possible existence of various inhomogeneous charge and spin states such as lattice and magnetic polarons, droplet and stripe structures, etc. [Dagotto et al., 2001; Nagaev, 2001; Kagan and Kugel, 2001]. Analogous phenomena are well known for many strongly correlated systems where the electron-electron interaction energy is higher than the kinetic energy. One of the most spectacular manifestations of such a behavior, i.e. the formation of ferromagnetic (FM) droplets

S. Halilov (ed.), Physics of Spin in Solids: Materials, Methods and Applications, 177–194.

(ferrons) was predicted in Ref. [Nagaev, 1967] for low-doped antiferromagnetic (AFM) semiconductors. Another example is a formation of a string (linear trace of frustrated spins) upon the motion of a hole in an AFM isolator [Bulaevskii et al., 1968]. Both these examples refer to the so-called electron phase separation, when a single charge carrier changes locally its electronic environment. In addition to this nanoscale phase separation, manganites can also exhibit a large-scale phase separation corresponding to the coexistence of different phases characteristic of first-order phase transitions (e.g., the transition between AFM and FM states). An example of this large-scale phase separation is given by the formation of relatively large FM droplets inside the AFM matrix. These droplets with linear sizes of about 100-1000 Å were observed in several experiments, in particular, by neutron diffraction methods in Ref. [Balagurov et al., 2001]. Note also that the attraction between one-electron ferromagnetic droplets (mediated by either elastic or magneto-dipole interaction) can result in merging of the ferrons and formation of intermediate to large-scale inhomogeneities [Lorenzana et al., 2001]. There exist clear experimental indications suggesting that the phase separation is inherent for both magnetically ordered phases and the paramagnetic state [Dagotto et al., 2001; Nagaev, 2001; Kagan and Kugel, 2001; Solin et al., 2003]. Therefore, the formation of inhomogeneous states proved to be a typical phenomenon for manganites in different parts of their phase diagram. Moreover, the phase separation should strongly affect the magnetic and transport properties of manganites.

Phase separation arguments are most often used for the domain of the existence of antiferromagnetism and especially in the vicinity of a transition between AFM and FM states. However, as we mentioned earlier, a manganite can be inhomogeneous even in the paramagnetic state at temperatures exceeding the corresponding phase transition temperature. An analysis of experimental data reveals a substantial similarity in the high-temperature behavior of resistivity, magnetoresistance, and magnetic susceptibility for various manganites with different low-temperature states [Babushkina et al., 2003; Fisher et al., 2003; Wagner et al., 2002; Zhao et al., 2001]. In addition, the magnetoresistance turns out to be rather large far from the FM-AFM transition and even in the paramagnetic region. Furthermore, the magnetic susceptibility of manganites is substantially higher than that for typical antiferromagnets. These experimental data clearly suggest the existence of significant FM correlations in the high-temperature range.

Here, we start from the assumption that the ferromagnetically correlated regions exist in manganites above the temperatures characterizing the onset of the long-range magnetic (FM or AFM) ordering. This assumption allows us to describe the characteristic features of resistivity, magnetoresistance, and magnetic susceptibility of manganites in the non-metallic state within the framework of one model. Below, we base our discussion on the model of conductivity of phase-separated manganites developed in Ref. [Babushkina et al., 2003; Rakhmanov et al., 2001; Sboychakov et al., 2002; Sboychakov et al., 2003] and use experimental data for manganites of different compositions reported in Ref. [Babushkina et al., 2003; Fisher et al., 2003; Wagner et al., 2002; Zhao et al., 2001]. Note that in this paper we do not limit ourselves by consideration of only one-electron magnetic droplets (ferrons) but rather generalize previously obtained results to the case of arbitrary number of electrons in ferromagnetically correlated domains.

In Section 2, the temperature dependence of resistivity is analyzed for the inhomogeneous state with the density of FM-correlated regions being far from the percolation threshold. In Sections 3 and 4, within the same assumptions, we discuss the magnetoresistance of manganites and their magnetic susceptibility, respectively. As a result, it is shown that the model of inhomogeneous state provides a good description for the high-temperature behavior of manganites.The comparison of theoretical results and experimental data allows us to reveal the general characteristics of ferromagnetically correlated regions.

12.2 Resistivity

In the analysis of the temperature dependence of resistivity, we will have in mind the physical picture discussed in the paper [Rakhmanov et al., 2001]. That is, we consider a non-ferromagnetic insulating matrix with small ferromagnetic droplets embedded in it. Charge transfer occurs via tunnelling of charge carriers from one droplet to another. A tunnelling probability depends, strictly speaking, upon applied magnetic field. We assume that the droplets do not overlap and the whole system is far from the percolation threshold. Each droplet can contain k charge carriers. When a new charge carrier tunnel to a droplet, it encounters with the Coulomb repulsion from the carriers already residing at this droplet. The repulsion energy A is assumed to be relatively large ($A > k_BT$). In this case, the main contribution to the conductivity is related to the processes involving the droplets containing k, $k+1$, or $k-1$ carriers. The corresponding expression for the resistivity $\rho(T)$ has the form

$$\rho = \frac{k_BT\exp(A/2k_BT)}{128\pi e^2\omega_0 l^5 kn^2}, \tag{1}$$

where e is the charge of the electron, ω_0 determines the characteristic energy of electrons in a droplet, l is the characteristic tunnelling length, and

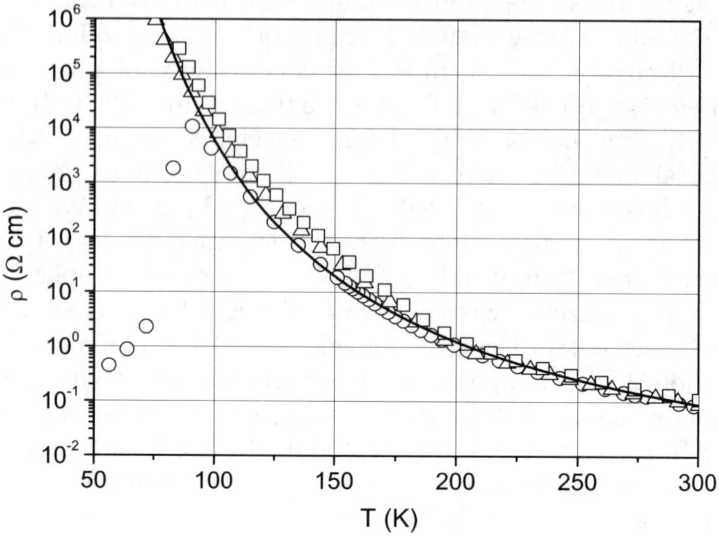

Figure 1. Temperature dependence of the resistivity for $(La_{1-y}Pr_y)_{0.7}Ca_{0.3}MnO_3$ samples [Babushkina et al., 2003]. Squares, triangles, and circles correspond to $y = 1$ (with $^{16}O \rightarrow {}^{18}O$ isotope substitution), $y = 0.75$ (with $^{16}O \rightarrow {}^{18}O$ isotope substitution), and $y = 0.75$ (with ^{16}O), respectively. Solid line is the fit based on Eq. (1).

n is the concentration of ferromagnetic droplets. Expression (1) could be easily derived by the method described in Ref. [Rakhmanov et al., 2001]. This expression is a straightforward generalization of the corresponding formula for the conductivity obtained for the case of one-electron droplets [Rakhmanov et al., 2001]. Electrical resistivity (1) exhibits a thermoactivation behavior where activation energy is equal to one half of the Coulomb repulsion energy (for details see Ref. [Rakhmanov et al., 2001]).

Expression (1) provides a fairly good description for the temperature dependence of the electrical resistivity for various manganites. As an illustration, in Figs. 1-4, we present experimental $\rho(T)$ curves for six different materials. Experimental data are plotted for samples reported in Ref. [Babushkina et al., 2003; Fisher et al., 2003; Wagner et al., 2002; Zhao et al., 2001]. The authors of these papers kindly provided us by the detailed numerical data on their measurements. As it could be seen from the figures and their captions, the examined samples differ in their chemical composition, type of crystal structure, magnitude of electrical resistivity (at fixed temperature, the latter varies for different

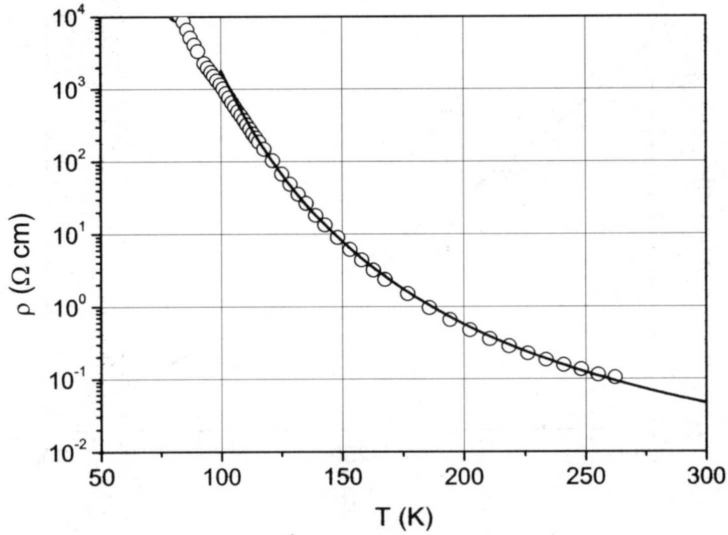

Figure 2. Temperature dependence of the resistivity for $Pr_{0.71}Ca_{0.29}MnO_3$ sample [Fisher et al., 2003]: experimental data (circles) and theoretical curve (solid line) based on Eq. (1).

samples by more than two orders of magnitude), and also by their low-temperature behavior (which is metallic for some samples and insulating for the others). On the other hand, in the high-temperature range (above the point of ferromagnetic phase transition), $\rho(T)$ exhibits a similar behavior for all the samples, which is well fitted by the relationship $\rho(T) \propto T \exp(A/2k_BT)$ (solid lines in the figures).

Based on Eq. (1) and experimental data, one can deduce some quantitative characteristics of the phase-separated state. In particular, the analysis carried out in the papers [Zhao et al., 2001; Zhao et al., 2002] demonstrated that an accurate estimate for the value of Coulomb energy A can be found by fitting experimental data and using Eq. (1). The data represented in Fig. 1-4 suggest that the Coulomb barrier A can be determined with an accuracy of 2-3% and its value lies in the narrow range from 3500 to 3700 K (see Table 1). As it was mentioned in the papers [Zhao et al., 2001; Rakhmanov et al., 2001; Zhao et al., 2002], the characteristic frequency ω_0 in (1) can also vary in a restricted range of 10^{13}-10^{14} Hz. This estimate might be derived, for example, from the uncertainty principle: $\hbar\omega_0 \sim \hbar^2/2ma^2$, where a is a characteristic droplet size, and m is the electron mass. Assuming $a \sim 1 - 2\,\text{nm}$, one

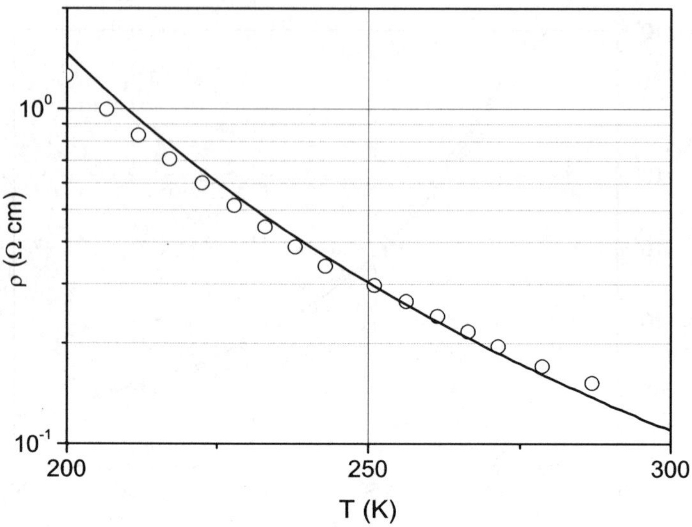

Figure 3. Temperature dependence of the resistivity for a layered manganite $(La_{0.4}Pr_{0.6})_{1.2}Sr_{1.8}Mn_2O_7$ [Wagner et al., 2002]: experimental data (circles) and theoretical curve (solid line) based on Eq. (1).

obtains the latter estimate. Note also that these values of a droplet size allow us to find an estimate for the barrier energy A, which is accurate within the order of magnitude. This energy is of the order of $e^2/\varepsilon a$, and substituting permittivity $\varepsilon \sim 10$, we get a value of A consistent with the experimental data.

Table 1.

Samples	A, K	ρ(200 K), Ω·cm	l^5n^2k, cm^{-1}	*Data source*
$(La_{1-y}Pr_y)_{0.7}Ca_{0.3}MnO_3$	3650	1.25	$2 \cdot 10^5$	Fig. 1 [a)]
$Pr_{0.71}Ca_{0.29}MnO_3$	3500	0.57	$3 \cdot 10^5$	Fig. 2 [b)]
$(La_{0.4}Pr_{0.6})_{1.2}Sr_{1.8}Mn_2O_7$ [*)]	3600	1.5	$1.5 \cdot 10^5$	Fig. 3 [c)]
$La_{0.8}Mg_{0.2}MnO_3$	3700	283	$1 \cdot 10^3$	Fig. 4 [d)]

a) [Babushkina et al., 2003]
b) [Fisher et al., 2003]
c) [Wagner et al., 2002]
d) [Zhao et al., 2001]
*) The chemical formula of this composition can be written as $(La_{0.4}Pr_{0.6})_{2-2x}Sr_{1+2x}Mn_2O_7$

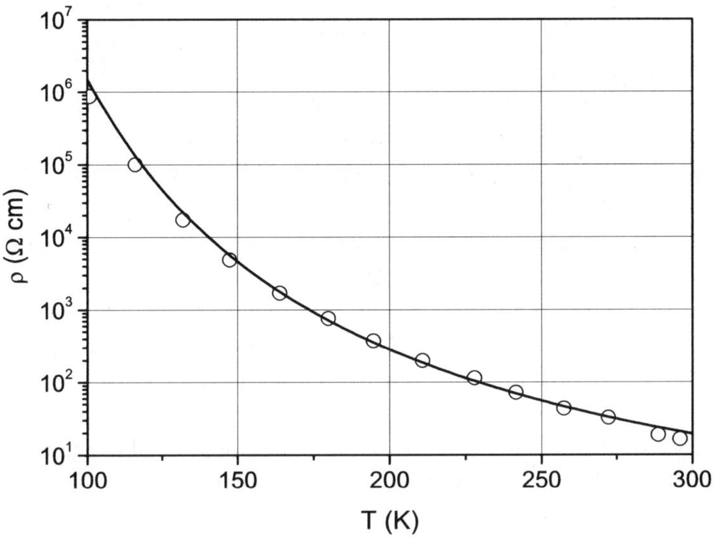

Figure 4. Temperature dependence of the resistivity for $La_{0.8}Mg_{0.2}MnO_3$ sample [Zhao et al., 2001]: experimental data (circles) and theoretical curve (solid line) based on Eq. (1).

It is rather difficult to estimate the tunnelling length l. However, we can say that in the domain of the applicability of relationship (1), length l cannot be much smaller than an interdroplet spacing [Rakhmanov et al., 2001]. In another situation, the behavior of the resistivity would be different. In the quasiclassical approximation, the tunnelling length is of the order of the characteristic size for the wave function provided the barrier height is comparable with the depth of the potential well. In our case, the size of the electron wave function is of the order of a ferron size, while the height of the barrier practically coincides with the depth of the potential well. The latter naturally follows from the model of ferron formation [Nagaev, 2001]. Therefore, it seems reasonable to assume the tunnelling length to be of the same order as a ferron size (few nanometers), though, generally speaking, it can substantially differ from a.

It is rather nontrivial task to estimate the concentration n of ferrons. In fact, following the papers [Zhao et al., 2001; Zhao et al., 2002], concentration n could be determined by the dopant concentration x as $n \approx x/d^3$. Yet this approach would bring at least two contradictions. First, even under the moderate concentration of divalent element

$x = 0.1 - 0.2$ the droplets should overlap giving rise to the continuous metallic and ferromagnetic cluster. However, the material could be insulating even at larger concentrations ($x = 0.5 - 0.6$), at least, in a high-temperature range. Second, as it can be seen from the experimental data, the relation between a dopant concentration and the conductivity of manganites is relatively complicated - for some materials changing x by a factor of two can change resistivity by two orders of magnitude [Zhao et al., 2001; Zhao et al., 2002], for other materials $\rho(x)$ exhibits even a nonmonotonic behavior in certain concentration ranges. Note that these discrepancies are essential not only for our model of phase separation but also for other models dealing with the properties of manganites (e.g., polaronic models [Ziese and Srinitiwarawong, 1998; Jakob et al., 1998]). Unfortunately, the authors of the papers [Zhao et al., 2001; Zhao et al., 2002] do not take into account these considerations when analyzing their results from the standpoint of the existing theories of the conductivity in manganites. The natural conclusion is that the number of carriers, which contribute to the charge transfer processes does not coincide with the concentration of the divalent dopant x. This is particularly obvious in the case of charge ordering when some part of the carriers introduced by doping becomes localized and forms a regular structure.

Therefore, using expression (1) and experimental data, we are able to obtain also the value of the combination $l^5 n^2 k$. In Table 1, the values of Coulomb energy A, resistivity ρ at 200 K and, combination $l^5 n^2 k$ are presented. All estimations were made based on Eq. (1) and the experimental data of Fig. 1-4. Note that whereas the accuracy of the estimate for A is about ± 50 K, the combination $l^5 n^2 k$ could be estimated only by the order of magnitude (at least, due to the uncertainty in the values of frequency ω_0).

12.3 Magnetoresistance

In the papers [Babushkina et al., 2003; Sboychakov et al., 2002; Sboychakov et al., 2003], it was demonstrated that the model of phase separation considered here results in a rather specific dependence of the magnetoresistance $MR(T, H)$ on temperature and magnetic field. At relatively high temperatures and not very strong magnetic fields, the expression for the magnetoresistance reads

$$MR \approx 5 \cdot 10^{-3} \frac{\mu_B^3 S^5 N_{ef}^3 Z^2 g^3 J^2 H_a}{(k_B T)^5} H^2, \qquad (2)$$

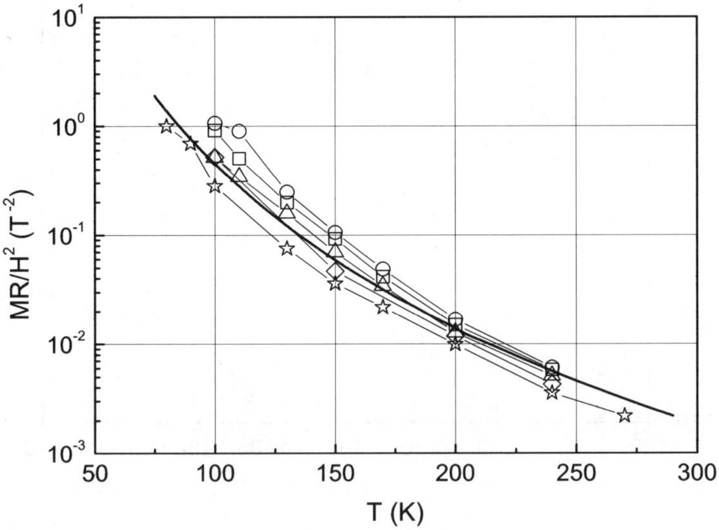

Figure 5. Temperature dependence of MR/H^2 ratio for $(La_{1-y}Pr_y)_{0.7}Ca_{0.3}MnO_3$ samples [Babushkina et al., 2003]. Squares, triangles, circles, diamonds, and asterisks correspond to $y = 0.75$, $y = 0.75$ (with 30% of ^{18}O), $y = 0.75$ (with $^{16}O \to {}^{18}O$ isotope substitution), $y = 1$, and $y = 1$ (with $^{16}O \to {}^{18}O$ isotope substitution), respectively. Solid line is the fit based on Eq. (2) ($MR \propto 1/T^5$).

where μ_B is the Bohr magneton, S is the average spin of a manganese ion, N_{ef} is the number of manganese atoms in a droplet, Z is the number of nearest neighbors of a manganese ion, g is the Landé factor, J is the exchange integral of the ferromagnetic interaction, and H_a is the effective field of magnetic anisotropy of a droplet. The $MR \propto H^2/T^5$ dependence was observed in the experiments for a number of manganites in the region of their non-metallic behavior [Babushkina et al., 2003; Fisher et al., 2003]. The same high-temperature behavior of the magnetoresistance can be obtained by processing the experimental data reported in Ref. [Wagner et al., 2002; Zhao et al., 2001] (see Figs. 5-8).

The value of S depends on the relative content of a trivalent and a tetravalent manganese ions and ranges from 3/2 to 2. Below it is assumed that $S = 2$ for all the estimations. Parameter Z is, in fact, the number of manganese ions interacting with a conduction electron placed in a droplet. It is reasonable to assume that Z is of the order of the number of nearest-neighbor sites around a manganese ion, i.e.

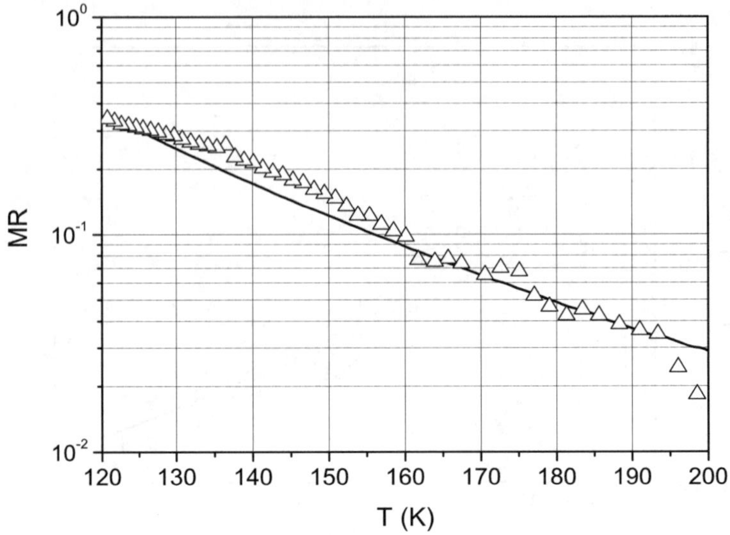

Figure 6. Temperature dependence of the magnetoresistance for $Pr_{0.71}Ca_{0.29}MnO_3$ sample at $H = 2T$: experimental data (triangles) [Fisher et al., 2003] and theoretical curve (solid line) based on Eq. (2).

$Z \approx 6$. The Landé factor g is determined from the experimental data. For manganese, g is usually assumed to be close to its spin value 2. The exchange integral J characterizes the magnetic interaction between a conduction electron and the molecular field generated by ferromagnetically correlated spins in a droplet. It is this molecular field that produces

Table 2.

Samples	N_{ef}	x	k	*Data source*
$(La_{1-y}Pr_y)_{0.7}Ca_{0.3}MnO_3$	250	0.3	75	Fig. 5 [a)]
$Pr_{0.71}Ca_{0.29}MnO_3$	200	0.29	58	Fig. 6 [b)]
$(La_{0.4}Pr_{0.6})_{1.2}Sr_{1.8}Mn_2O_7$ [*)]	250	0.4	100	Fig. 7 [c)]
$La_{0.8}Mg_{0.2}MnO_3$	265	0.2	53	Fig. 8 [d)]

[a)] [Babushkina et al., 2003]
[b)] [Fisher et al., 2003]
[c)] [Wagner et al., 2002]
[d)] [Zhao et al., 2001]
[*)] The chemical formula of this composition can be written as $(La_{0.4}Pr_{0.6})_{2-2x}Sr_{1+2x}Mn_2O_7$

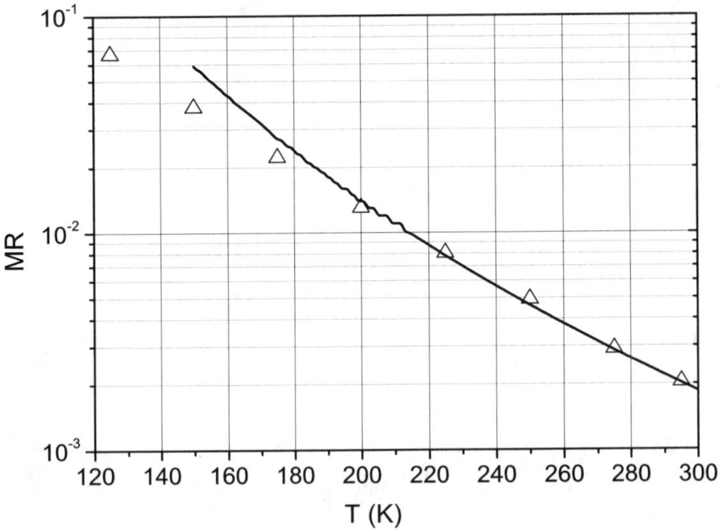

Figure 7. Temperature dependence of the magnetoresistance for $(La_{0.4}Pr_{0.6})_{1.2}Sr_{1.8}Mn_2O_7$ sample at $H = 1T$: experimental data (triangles) [Wagner et al., 2002] and theoretical curve (solid line) based on Eq. (2).

a ferromagnetic state at low temperatures. Therefore, we can use a well-known relationship $S(S+1)ZJ/3 = k_B T_C$ of the molecular field theory to evaluate the exchange integral (here T_C is the Curie temperature). The value of T_C is determined from the experiment (based on neutron diffraction or magnetization measurements). For example, in La-Pr-Ca manganites, it is about 100 – 120 K [Balagurov et al., 2001].

The magnetic anisotropy of manganites related to crystal structure of these compounds is usually not too high. This implies that the main contribution to the effective field of a magnetic anisotropy H_a stems from the shape anisotropy of a droplet and can be evaluated as $H_a = \pi(1 - 3\tilde{N})M_s$, where $\tilde{N}$ is the demagnetization factor of the droplet (along the main axis), M_s is the magnetic moment per unit volume of the droplet. Below we assume a droplet to be sufficiently elongated ($\tilde{N} \ll 1$) and $M_s = Sg\mu_B/d^3$. Then $H_a \approx 2$ kOe.

The value of N_{ef} is determined by the size of a droplet and it could be found from the neutron diffraction experiments. However, we are unaware of such measurements performed for the systems under discussion in a wide temperature range. Therefore, N_{ef} is treated here as a fitting parameter. Hence, using Eq. (2) and the above estimates, we can deter-

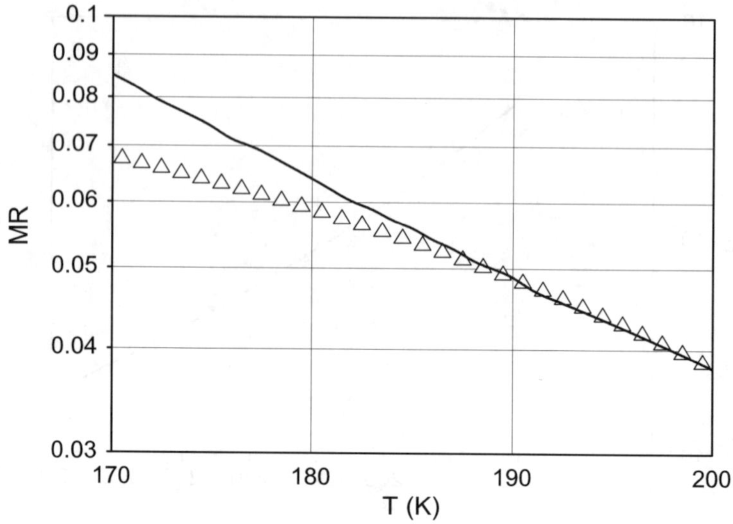

Figure 8. Temperature dependence of the magnetoresistance for $La_{0.8}Mg_{0.2}MnO_3$ sample at $H = 1.5$T: experimental data (triangles) [Zhao et al., 2001] and theoretical curve (solid line) based on Eq. (2).

mine the value of N_{ef} from the experimental data on the magnetoresistance (in the range of parameters corresponding to $MR \propto H^2/T^5$). The results are summarized in Table 2. In Figs. 5-8, solid curves correspond to the fitting procedure based on Eq. (2). The value of T_C was chosen to be equal to 120 K.

As a result, the size of the ferromagnetically correlated regions turns out to be nearly the same at temperatures about 200-300 K for all compositions under discussion. The volume of these regions is approximately equal to that of a ball with 7-8 lattice constants in diameter. It is natural to assume that within a droplet the number of charge carriers contributing to tunnelling processes equals to the number of dopant atoms. Hence, we can write that $k = N_{ef}x$, where x is the atomic percentage of dopants. The values of x and k are presented in Table 2.

12.4 Magnetic susceptibility

The concentration of droplets can be evaluated based on the magnetic susceptibility data, if we assume that the dominant contribution to the susceptibility comes from the ferromagnetically correlated regions. At high temperatures ($k_BT \gg \mu_B gSN_{ef}H, \mu_B gSN_{ef}H_a$), susceptibility

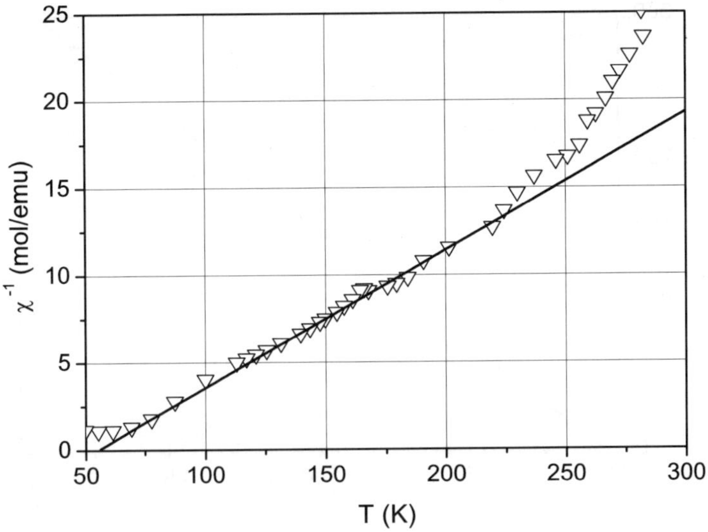

Figure 9. Temperature dependence of the inverse magnetic susceptibility for $La_{1-y}Pr_y)_{0.7}Ca_{0.3}MnO_3$ sample at $y = 1$: experimental data (triangles) [Babushkina et al., 2003] and theoretical curve (solid line) based on Eq. (3). For the other samples of this group, the behavior of $\chi(T)$ at high temperatures is rather similar to that illustrated in this figure (see Ref. [Babushkina et al., 2003]).

$\chi(T)$ can be written as

$$\chi(T) = \frac{n(\mu_B g S N_{ef})^2}{3k_B(T - \Theta)}, \tag{3}$$

Table 3.

Samples	Θ, K	n, cm^{-3}	p	l, Å	*Data source*
$(La_{1-y}Pr_y)_{0.7}Ca_{0.3}MnO_3$	55	$1.8 \cdot 10^{18}$	0.03	24	Fig. 9 [a)]
$Pr_{0.71}Ca_{0.29}MnO_3$	105	$6.0 \cdot 10^{18}$	0.07	17	Fig. 10 [b)]
$(La_{0.4}Pr_{0.6})_{1.2}Sr_{1.8}Mn_2O_7$ [*)]	255	$2.5 \cdot 10^{18}$	0.04	19	Fig. 11 [c)]
$La_{0.8}Mg_{0.2}MnO_3$	150	$0.6 \cdot 10^{18}$	0.01	14	Fig. 12 [d)]

[a)] [Babushkina et al., 2003]
[b)] [Fisher et al., 2003]
[c)] [Wagner et al., 2002]
[d)] [Zhao et al., 2001]
[*)] The chemical formula of this composition can be written as $(La_{0.4}Pr_{0.6})_{2-2x}Sr_{1+2x}Mn_2O_7$

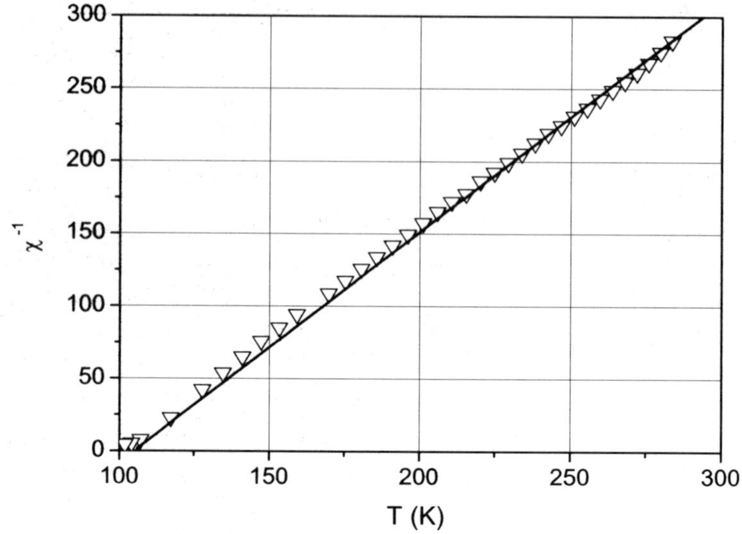

Figure 10. Temperature dependence of the inverse magnetic susceptibility for $Pr_{0.71}Ca_{0.29}MnO_3$ sample: experimental data (triangles) [Fisher et al., 2003] and theoretical curve (solid line) based on Eq. (3). The sample was porous, its density was assumed to differ by a factor of 0.7 from the theoretical value.

where Θ is the Curie-Weiss constant. The results of the processing of the experimental data are presented in Table 3. In Figs. 9-12, the solid curves correspond to the fitting procedure based on Eq. (3). Using these results, we can also estimate the concentration of ferromagnetic phase as $p = nN_{ef}d^3$. For all the samples, the value of the lattice constant d was taken to be equal to 3.9 Å. Based on the data of Tables 1-3, it is also possible to find an estimate for the tunnelling length l.

12.5 Discussion

To sum up, the analysis performed in the previous sections demonstrates that a simple model of the electron tunnelling between the ferromagnetically correlated regions (FM droplets) provides a possibility to describe the conductivity and the magnetoresistance data for a wide class of manganites. The comparison of the theoretical predictions with the experimental data on the temperature dependence of the resistivity, magnetoresistance, and magnetic susceptibility enables us to reveal various characteristics of the phase-separated state such as the size of

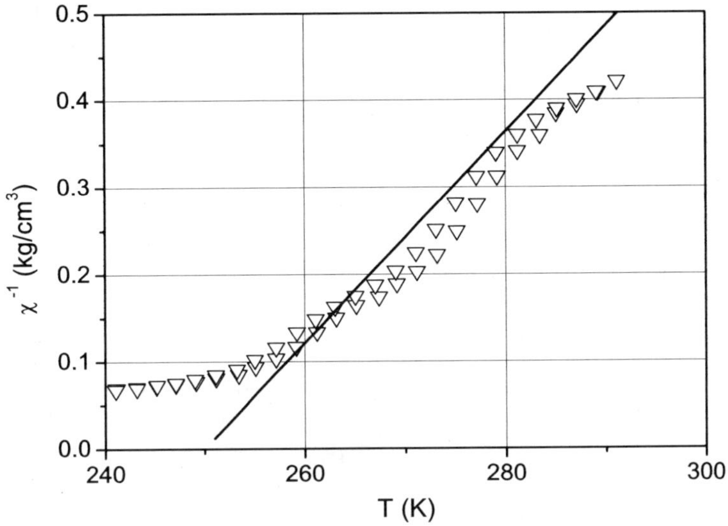

Figure 11. Temperature dependence of the inverse magnetic susceptibility for the sample of $(La_{0.4}Pr_{0.6})_{1.2}Sr_{1.8}Mn_2O_7$layered manganite: experimental data (triangles) [Wagner et al., 2002] and theoretical curve (solid line) based on Eq. (3).

FM droplets, their density, the number of electrons in a droplet and also to estimate the characteristic tunnelling length of the charge carriers. The determined values of parameters appear to be rather reasonable. Indeed, the characteristic tunnelling length turns out to be of the order of FM droplet size, the concentration of the ferromagnetic phase in the high-temperature range is substantially smaller than the percolation threshold and varies from about 1 % to 7 %.

Note also that the droplets contain 50-100 charge carriers, whereas parameter A deduced from the experimental data is equal by the order of magnitude to the energy of Coulomb repulsion in a metallic ball of $(7 \div 8)\,d$ in diameter. The obtained numerical values for the droplet parameters (characteristic tunnelling barrier, size, and tunnelling length) are close for manganites with drastically different transport properties.

The large magnitude of the $1/f$ noise in the temperature range corresponding to the insulating state is another characteristic feature of the phase-separated manganites [Podzorov et al., 2000; Podzorov et al., 2001]. In the framework of the model of phase separation discussed here, the following expression for the Hooge constant was derived in the

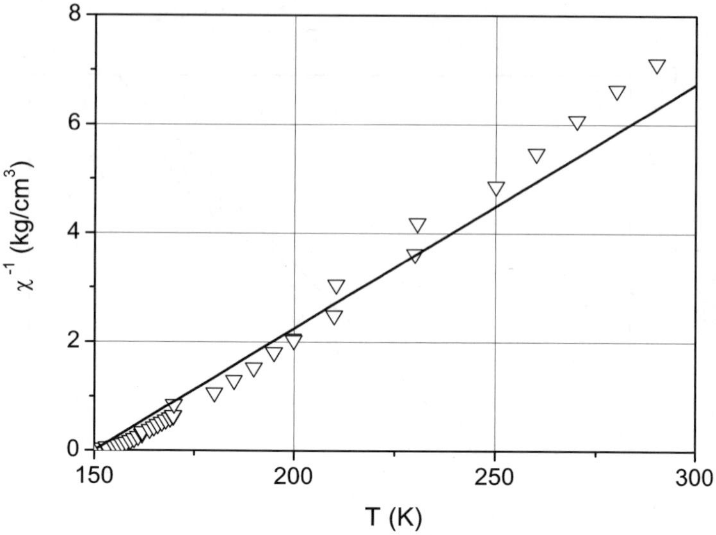

Figure 12. Temperature dependence of the inverse magnetic susceptibility for $La_{0.8}Mg_{0.2}MnO_3$ sample: experimental data (triangles) [Zhao et al., 2001] and theoretical curve (solid line) based on Eq. (3).

papers [Rakhmanov et al., 2001; Sboychakov et al., 2002]

$$\alpha_H = \frac{\langle \delta U^2 \rangle_\omega V_s \omega}{U_{DC}^2} = 2\pi^2 l^3 \ln^2 \left(\frac{\tilde{\omega}_0}{\omega} \right), \tag{4}$$

where $\langle \delta U^2 \rangle_\omega$ is the spectral density of the voltage fluctuations, V_s is the volume of a sample, U_{DC} is the applied voltage, and $\tilde{\omega}_0 = \omega_0 \exp(A/2k_BT)$. Substituting to Eq. (4) the estimated values of the parameters presented in the tables and in the text, we get $\alpha_H \approx 10^{-16}\,\mathrm{cm}^3$ at temperatures 100-200 K and frequencies 1-1000 s^{-1}. This value of α_H is by 3-5 orders of magnitude higher than the corresponding values for semiconductors.

Thus, we have a rather consistent scheme describing the transport properties of manganites under condition that the ferromagnetically correlated regions do not form a percolation cluster. Moreover, the presented approach proves to be valid for a fairly wide range of the dopant concentrations. However, as it was mentioned above, the relation between the concentration of ferromagnetic droplets and the doping level is far from being well understood. If the picture of the phase separation is believed to be applicable, it becomes obvious that not all electrons or holes introduced by doping participate in the transport processes. Below we try to present some qualitative arguments illustrating the possible difference in the effective concentration of charge carriers below and above the transition from paramagnetic to magnetically ordered state.

In the phase diagram of a typical manganite, one would have the AFM state with FM-phase inclusions in the low-temperature range and at a low doping level. The transition from AFM to FM phase occurs upon doping. At high temperatures, manganites are in the paramagnetic (PM) state. When the temperature decreases, we observe the transition from PM to AFM or FM state depending on the doping level.

Let us consider the behavior of such a system in the vicinity of a triple point. In the AFM phase, radius R of a region which one electron converts into FM state can be estimated as $R = d\left(\pi t/4J_{ff}S^2Z\right)^{1/5}$ [Kagan and Kugel, 2001], where J_{ff} is an AFM interaction constant. For high-temperature PM phase, a radius R_T of a region that one electron converts into FM state corresponds to the size of the so-called temperature ferron and equals to $R_T = d\left(\pi t/4k_BT\ln(2S+1)\right)^{1/5}$ [Kagan and Kugel, 2001]. The critical concentration $x_c \approx 0.15$ of the overlapping of low-temperature ferrons can be derived from the estimate $x_c \approx 3/4\pi \cdot (d/R)^3$, while for the high-temperature ferrons it follows from the estimate $\delta_c \approx 3/4\pi \cdot (d/R_T)^3$. Substituting the expressions for the radii of the high- and the low-temperature ferrons to the ratio x_c/δ_c, we obtain the following estimate for this ratio in the vicinity of the triple point corresponding to the coexistence of FM, AFM, and PM phases:

$$\frac{x_c}{\delta_c} \sim \left[\frac{T\ln(2S+1)}{zJ_{ff}S^2}\right]^{3/5} \sim \left[\frac{T_C\ln(2S+1)}{T_N}\right]^{3/5}, \tag{5}$$

where T_C and T_N are the Curie and the Neel temperatures, respectively. For the manganites under discussion, we have $T_C \sim T_N \sim$ 120-150 K and $\ln(2S+1) \sim 1.6$ for $S = 2$, hence $\delta_c \leq x_c$. The sign of this inequality is in agreement with experimental data which imply $\delta \sim 1-7\,\%$. Thus, we do not have a clear explanation of the charge disbalance in paramagnetic region in spite of the fact that the trend is correctly caught by our simple estimates. Probably, at $x > x_c$ (in real experiments the concentration x can be as high as 50 %), the residual charge is localized in the paramagnetic matrix outside the temperature ferrons. The detailed study of this problem will be presented elsewhere.

Acknowledgments

The authors are grateful to V.A. Aksenov, N.A. Babushkina, S.W. Cheong, I. Gordon, L.M. Fisher, D.I. Khomskii, F.V. Kusmartsev, V.V. Moshchalkov, A.N. Taldenkov, I.F. Voloshin, G. Williams and X.Z. Zhou for useful discussions and provided experimental data. This work was supported by the Russian Foundation for Basic Research (Grants Nos. 02-02-16708, 03-02-06320, and NSh-1694.2003.2), INTAS (Grant No. 01-2008), and CRDF (Grant No. RP2-2355-MO-02).

References

Babushkina, N. A., Chistotina, E. A., Kugel, K. I., Rakhmanov, A. L., Gorbenko, O. Yu., and Kaul, A. R. (2003). *J. Phys.: Condens. Matter*, 15:259.

Balagurov, A. M., Pomjakushin, V. Yu., Sheptyakov, D. V., Aksenov, V. L., Fischer, P., Keller, L., Gorbenko, O. Yu., Kaul, A. R., and Babushkina, N. A. (2001). *Phys. Rev. B*, 64:024420.

Bulaevskii, L. N., Nagaev, E. L., and Khomskii, D. I. (1968). *Zh. Eksp. Teor. Fiz.*, 54:1562. [*Sov. Phys. JETP*, 27:836].

Dagotto, E., Hotta, T., and Moreo, A. (2001). *Phys. Rep.*, 344:1.

Fisher, L. M., Kalinov, A. V., Voloshin, I. F., Babushkina, N. A., Khomskii, D. I., and Kugel, K. I. (2003). *Phys. Rev. B*, 68:174403.

Jakob, G., Westerburg, W., Martin, F., and Adrian, H. (1998). *Phys. Rev. B*, 58:14966.

Kagan, M. Yu. and Kugel, K. I. (2001). *Usp. Fiz. Nauk*, 171:577. [*Physics Uspekhi*, 44:553].

Lorenzana, J., Castellani, C., and Di Castro, C. (2001). *Phys. Rev. B*, 64:235128.

Nagaev, E. L. (1967). *Pis'ma Zh. Eksp. Teor. Fiz.*, 6:484. [*JETP Lett.*, 6:18].

Nagaev, E. L. (2001). *Phys. Rep.*, 346:387.

Podzorov, V., Uehara, M., Gershenson, M. E., and Cheong, S-W. (2001). *Phys. Rev. B*, 64:115113.

Podzorov, V., Uehara, M., Gershenson, M. E., Koo, T. Y., and Cheong, S-W. (2000). *Phys. Rev. B*, 61:R3784.

Rakhmanov, A. L., Kugel, K. I., Blanter, Ya. M., and Kagan, M. Yu. (2001). *Phys. Rev. B*, 63:174424.

Sboychakov, A. O., Rakhmanov, A. L., Kugel, K. I., Kagan, M. Yu., and Brodsky, I. V. (2002). *Zh. Eksp. Teor. Fiz.*, 122:869. [*JETP*, 95:753].

Sboychakov, A. O., Rakhmanov, A. L., Kugel, K. I., Kagan, M. Yu., and Brodsky, I. V. (2003). *J. Phys.: Condens. Matter*, 15:1705.

Solin, N. I., Mashkautsan, V. V., Korolev, A. V., Loshkareva, N. N., and Pinsard, L. (2003). *Pis'ma Zh. Eksp. Teor. Fiz.*, 77:275. [*JETP Lett.*, 77:230].

Wagner, P., Gordon, I., Moshchalkov, V. V., Bruynseraede, Y., Apostu, M., Suryanarayanan, R., and Revcolevschi, A. (2002). *Europhys. Lett.*, 58:285.

Zhao, J. H., Kunkel, H. P., Zhou, X. Z., and Williams, G. (2001). *J. Phys.: Condens. Matter*, 13:285.

Zhao, J. H., Kunkel, H. P., Zhou, X. Z., and Williams, G. (2002). *Phys. Rev. B*, 66:184428.

Ziese, M. and Srinitiwarawong, C. (1998). *Phys. Rev. B*, 58:11519.

NEW MAGNETIC SEMICONDUCTORS ON THE BASE OF TLB^{VI}-MEB^{VI} SYSTEMS (ME-FE, CO, NI, MN; B-S, SE, TE)

E. M. Kerimova, S. N. Mustafaeva, A. I. Jabbarly, G. Sultanov, A.I. Gasanov, R. N. Kerimov

Institute of Physics, National Academy of Sciences of Azerbaijan, Baku, Azerbaijan.

Abstract Fe-Tl, Co-Tl, Ni-Tl, and Mn-Tl chalcogenides are representatives of a new class of magnetic semiconductors. We propose a methods of synthesizing of TlMeB^{VI} samples and present the results of investigations of electrical, thermoelectrical and magnetic properties of the prepared compounds.

Keywords: Chalcogenides, magnetic semiconductors, thermoelectrical and magnetic properties.

Fe-Tl, Co-Tl, Ni-Tl, and Mn-Tl chalcogenides are representatives of new class of magnetic semiconductors. Some physical properties of these compounds were studied earlier in [1-7]. In the present work, we propose a methods of synthesizing of TlMeB^{VI} samples and present the results of investigations of electrical, thermoelectrical and magnetic properties of the prepared compounds.

The synthesis of $TlNiS_2$ was carried out in an ampule evacuated to pressure 10^{-3} Pa. The ampule was fabricated from a fused silica tube. In this case, $TlNiS_2$ samples were prepared through the interaction of initial elements (Tl, Ni, S) of high -purity grade. In order to prevent the ampule filled with reactants from explosion, the furnace temperature was raised to the melting temperature of sulfur (391K) and the ampule was held at this temperature for 3 h. Then, the furnace temperature was raised to 1400K at a rate of 100 K/h and the ampule was held at this temperature for 1.5-2.0 h, after which it was cooled to 300K. Thereafter the ampule was broken; the alloy contained in it was crushed to powder, the powder thus prepared was placed in a new ampule, which was then evacuated to a pressure of 10^{-3} Pa; and the above process was repeated

S. Halilov (ed.), Physics of Spin in Solids: Materials, Methods and Applications, 195–206.

with subsequent cooling to 600K. At this temperature, the $TlNiS_2$ sample was annealed for 240 h.

The $TlNiS_2$ samples thus synthesized were subjected to x-ray powder diffraction analysis on a DRON-3M diffractometer (Cu_{K_α} radiation, Ni filter; λ_α=1.5418Å). The x-ray diffraction patterns were recorded continuously. The diffraction angles were determined by measuring the intensity peaks. The error in determining the angles of reflections did not exceed 0,02° . For the $TlNiS_2$ sample, 24 diffraction reflections measured were unambiguously indexed in the hexagonal system with the lattice parameters: a=12.2754Å; c=19.3178A; z=32; ρ =6.896 g/cm^3.

20 diffraction reflections, fixed of $TlNiSe_2$ sample are displayed on the basis of tetragonal syngony with the lattice parameters: a=10.2015Å; c=20.8632Å; z=27; ρ=8.692 g/cm^3.

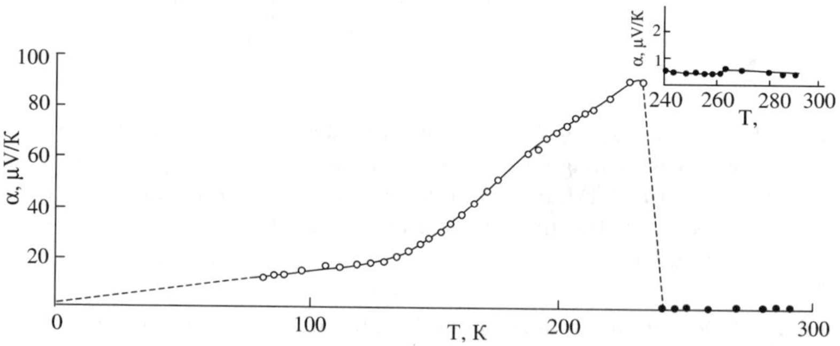

Figure 1. Temperature dependence of the thermopower in $TlNiS_2$. The inset shows the high-temperature branch α(T) on an enlarged scale.

The synthesis regims of $TlMnSe_2$ and $TlMnS_2$ phases have been worked out. X-ray analysis showed that $TlMnS_2$ is crystallized in tetragonal structure with elementary cell parameters: a=6.53; c=23.96A; z=8; ρ=6.71 g/cm^3.

Phase relations in TlSe-FeSe system were studied, and the compound $TlFeSe_2$ was identified. It was established that $TlFeSe_2$ is congruently melting compound (T_m=903K). $TlFeSe_2$ single crystals were obtained by Bridgmen - Stokberger method. XRD data indicate that $TlFeSe_2$ crystallizes in the monoclinic structure with lattice parameters: a=12.02Å; b=5.50Å; c=7.13A ; β=118.52°.

Phase relations in TlSe-CoSe system were studied too, and the compound $TlCoSe_2$ was identified. It was established that $TlCoSe_2$ is con-

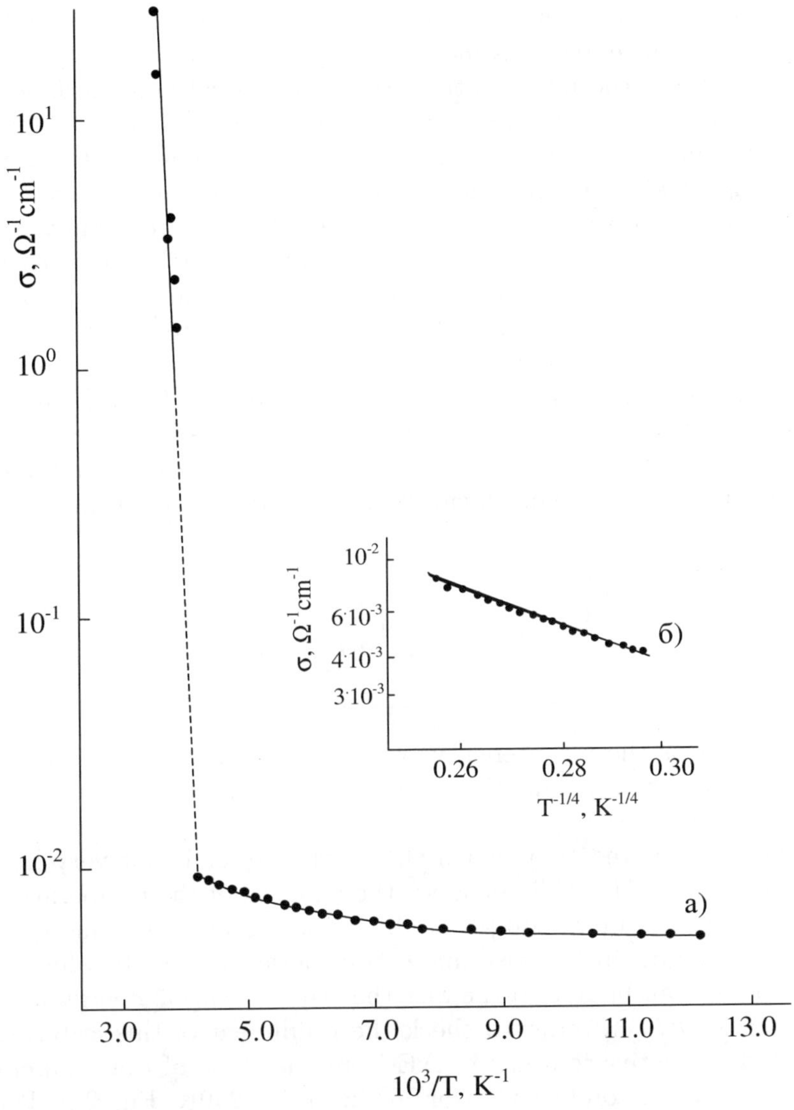

Figure 2. Dependence of the conductivity of $TlNiS_2$ on (a) $10^3/T$ and (b) $T^{-1/4}$.

gruently melting compound with T_m=650± 10K. XRD data indicate that $TlCoSe_2$ crystallizes in the hexagonal system with lattice parameters: a=3.747Å; c=22.772Å.

Complete phase diagram of TlTe-FeTe system was studied. It was established that liquidus curve of TlTe-FeTe system consist of crystallization regions of Tl_2Te, $TlFeTe_2$ and FeTe compounds. Simple eutectics of $(TlTe)_{0,4}(FeTe)_{0,6}$ composition is formed between $TlFeTe_2$ and FeTe compounds. This eutectic is melted at 813K.

The results of the investigations of the electrical and thermoelectric properties of $TlNiS_2$ samples are as follows. Figure 1 depicts the temperature dependence of the thermopower for $TlNiS_2$ in the temperature range 80-300 K. As the temperature increases from 80 K, the thermopower increases first moderately and then more rapidly and, at T=235 K, reaches a maximum (91μV/K). With a further increase in the temperature, the thermopower sharply decreases from 91 to ˜0.5 μV/K and then remains nearly constant to room temperature. In fig. 1, the inset shows the high-temperature branch of the thermopower on a tenfold enlarged scale of the ordinate axis. The positive sign of the thermopower indicates that holes are the majority charge carriers in $TlNiS_2$.

According to [8], the thermopower of chalcogenide semiconductors in the case of p-type conduction can be represented in the form:

$$\alpha(T) = -\frac{k}{e}\left[\frac{\Delta E}{kT} + \gamma\right] \tag{1}$$

where γkT is the mean energy transferred by holes, $\gamma \approx 1$, ΔÅ is the activation energy of conduction, k is the Boltzmann constant, and e is the elementary charge.

It should be noted that, when the thermopower is not very high (of the order of k/e=86 μV/K or less), the analysis of the temperature dependence α(T) is more complicated. If the material remains a p-type semiconductor (as in the case under consideration), small values of the thermopower can be due to the fact that the activation energy ΔE is of the order of kT. In order to check the fulfillment of this criterion, we estimated the activation energy ΔE from the slope of the temperature dependence of the conductivity for $TlNiS_2$ at T<240K (Fig. 2a). It is evident from Fig. 2a, that the temperature dependence of the conductivity has a variable slope.

For this reason, we estimated the activation energy ΔE in the temperature range 160-240K; as a result, the activation energy was found to be equal to 1.54x10^{-2}eV. For these temperatures, the values of kT were determined to be (1.38-2.00)x10^{-2} eV. In other words, the values of ΔE and kT for $TlNiS_2$ at low temperature are actually of the same order of magnitude, as is the case with metals. In metals, the current is transferred by charge carriers in the energy band whose thickness is of the order of kT in the vicinity of the Fermi energy (E_F). According to

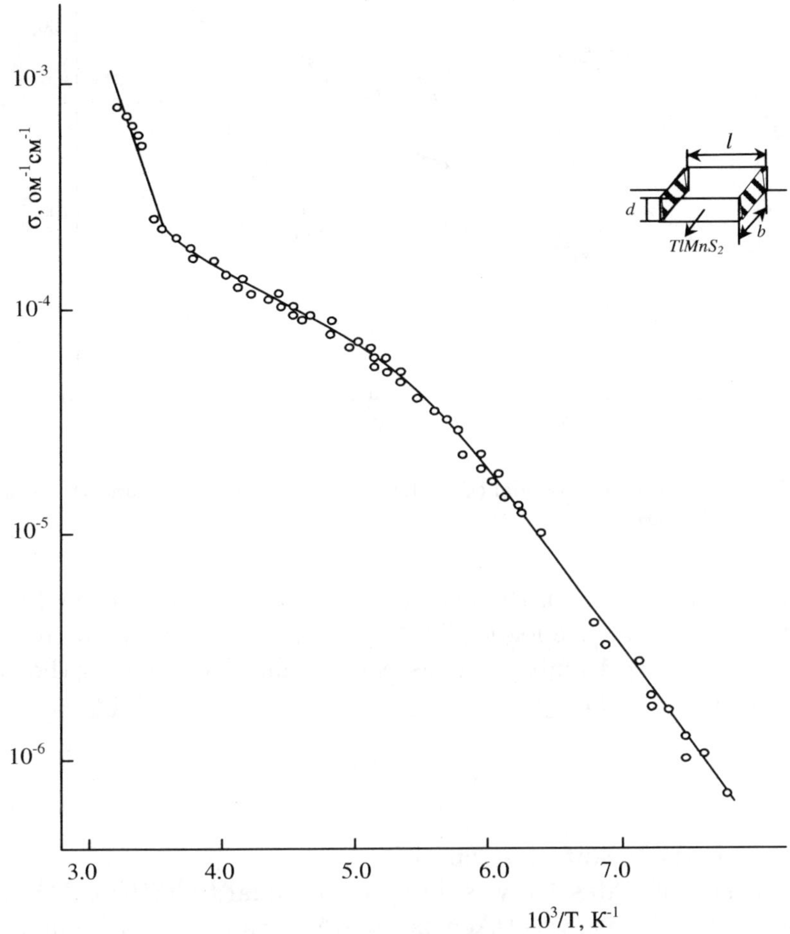

Figure 3. Temperature dependence of conductivity in $TlMnS_2$.

[8], the thermopower of a metal has the form

$$\alpha(T) = \frac{\pi^3}{3}\left(\frac{\partial \ln \sigma}{\partial E}\right)_{E=E_F} \tag{2}$$

Formula (2) is valid only when êÒ<< Å$_F$.

As was noted above and shown in Fig. 2a, the dependence of $\ln \sigma$ on 1/T at temperatures T<240K is characterized by a monotonic decrease in the activation energy with a decrease in the temperature. This behavior of the conductivity in $TlNiS_2$ at low temperatures suggests that charge transfer occurs through the variable-range-hopping mechanism [8], provided the current is transferred by charge carriers at the states

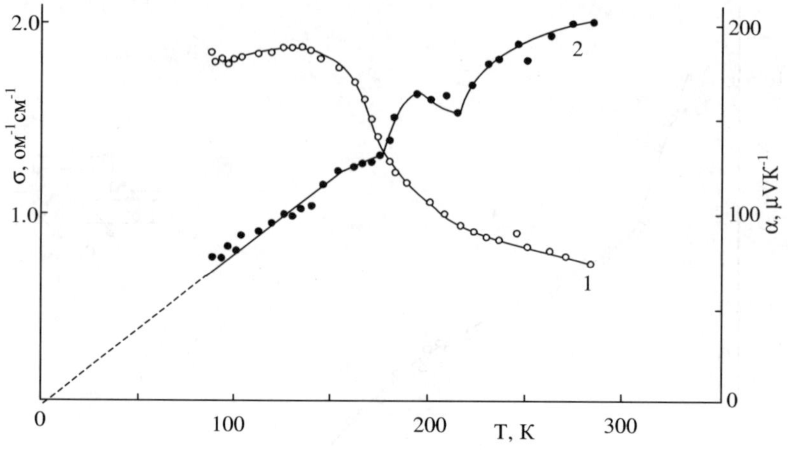

Figure 4. Temperature dependence of the conductivity (curve 1) and thermopower (curve 2) in $TlMnSe_2$.

localized in the vicinity of the Fermi level. This is also confirmed by the temperature dependence logσ ˜$T^{-1/4}$ (Fig. 2b). The slope of this curve (T_0) allowed us to estimate the density of localized states near the Fermi level from the formula [8]:

$$N_F = \frac{16}{T_0 k a^3} \tag{3}$$

where α is the localization length.

The density of states N_F was found to be equal to $9x10^{20}eV^{-1}cm^{-3}$. The localization length was taken as a=20Å (by analogy with binary sulfides of Group III elements [9]). Such a high value of N_F is characteristic of amorphous semiconductors. Therefore, it can be concluded that the energy-band structure of $TlNiS_2$ is similar to that of amorphous semiconductors. We calculated the hopping distance in $TlNiS_2$ according to the formula

$$R(T) = \frac{3}{8} a (T_0/T)^{1/4} \tag{4}$$

As a result, we found that, at T=110K, R≈30Å.

From the expression given in [8],

$$J = \frac{2}{2\pi R^3 N_F} \tag{5}$$

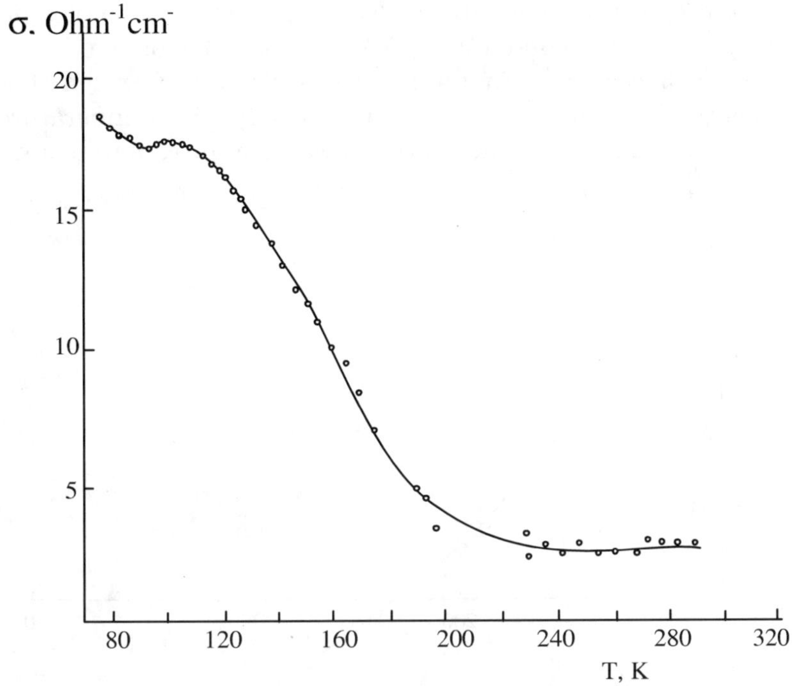

Figure 5. Temperature dependence of conductivity in $TlCoS_2$.

we estimated the scatter of the trapping states about the Fermi level: J $= 1.97\text{x}10^{-2}$ eV. As was shown above, the approximate activation energy of conduction ΔE, which was determined from the dependence $\ln\sigma$ on $10^3/T$ at a low temperatures, is of the same order of magnitude.

In the temperature range 80-110 K, the activation energy of conduction becomes zero. The activationless conduction also exhibits hopping nature, which manifests itself in the hopping charge carriers over spatially more distant but energetically more closely located centers without photon absorption [10].

In contrast to formula (2) for the thermopower of metals, the temperature dependence of α in the region of hopping conduction can be represented by the relationship [8]

$$\alpha(T) = A + BT, \tag{6}$$

where B is the temperature coefficient for the thermopower. In our case, the dependence α(T) for $TlNiS_2$ (Fig. 1) is characterized by two slopes.

In the temperature range 80-110 K, where $TlNiS_2$ samples possess activationless hopping conduction (ΔÅ=0), the slope of the curve α(T) is equal to 0.13 μV/ K^2. At temperatures T> 110 K, when the activation energy of conduction varies monotonically with temperature, the temperature coefficient for the thermopower is approximately six times larger: $B = \frac{\partial\alpha}{\partial T} = \mu V/\ K^2$.

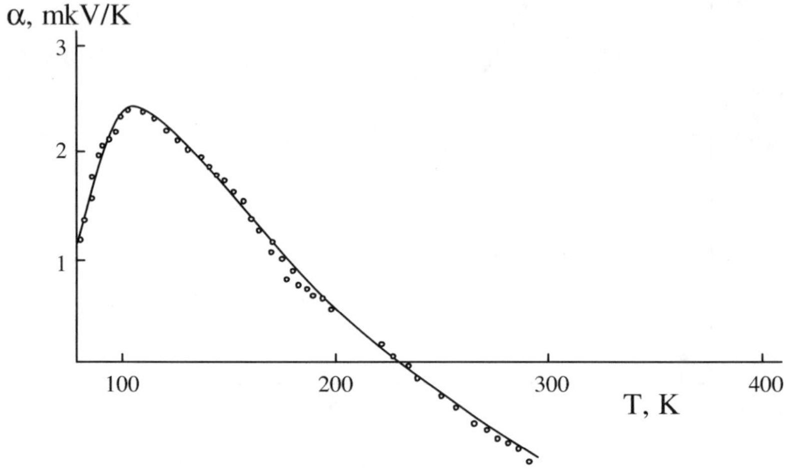

Figure 6. Dependence of thermo-e.m.f. on temperature in $TlCoS_2$.

The extrapolated low- temperature branch α(T) goes through zero, i.e., A=0 in formula (6). This indicates that, in the temperature range 80-110 K, where the electrical conductivity σ does not depend on T, the experimental values of α satisfy formula (2) for the thermopower of metals. In the temperature range 110-240 K, the thermopower obeys relationship (6). The thermopower is determined primarily by the density of states and, hence, has positive sign in the region of hopping conduction.

Earlier [1], we showed that similar behavior of α(T) is also observed in $TlFeSe_2$; i.e., under conditions of hopping conduction, the sign of the thermopower is positive and the temperature dependence α(T) is linear (α˜T).

It is evident from Figs. 1 and 2 that, at temperatures close to 240 K, the dependences α(T) and σ(T) exhibit a jump; i.e., the thermopower sharply decreases, whereas the conductivity increases by more than three orders of magnitude. In this range of temperatures, the slope of the curve $\ln\sigma$ (1/T) is estimated at ˜ 1.0eV. Such a sharp increase in the conductivity σ at the activation energy ΔE =1.0eV can be associated with the onset of intrinsic conduction. In the case when the current is transferred by carriers over states distributed throughout the whole sample, the parameter γ in formula (1) should be of the order of unity. At T≈240 K, the thermopower in $TlNiS_2$ sample, which was estimated from formula (1) at ΔE=1.0eV and γ =1, is more than one order of magnitude higher than the experimentally observed thermopower. In other words, the experimental values of α are not as large as those calculated with the activation energy ΔE obtained from the slope of the curve $\ln \sigma(1/T)$. Possibly, this deference is caused by the fact that, at high temperatures, both holes and electrons are involved in conduction. Of course, in this case, the thermopower α is less tan that calculated from formula (1), which holds for semiconductors with single-type charge carriers.

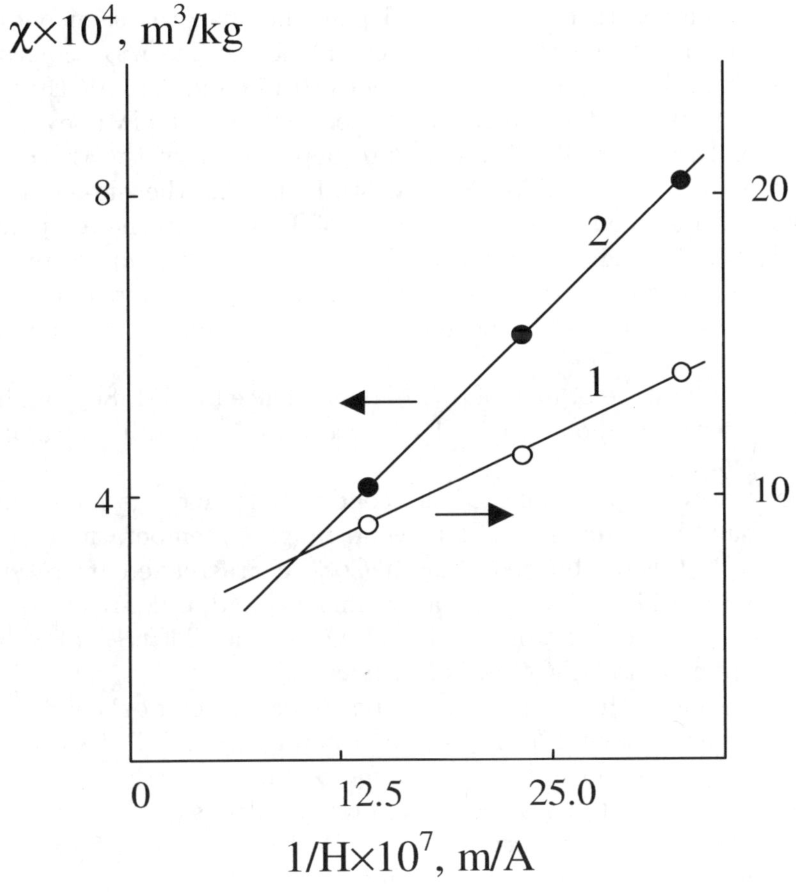

Figure 7. Temperature dependence of magnetic susceptibility of $TlFeSe_2$ at H=636 kA/m: 1 - H⊥c-axis ; 2 - H||*c*-axis of crystal.

Thus, it was demonstrated that, at low temperatures, when hopping conduction dominates, the thermopower of $TlNiS_2$ is proportional to the temperature. As the temperature increases, the charge carriers excited in the allowed band begin to dominate in conduction and the thermopower decreases drastically (by a factor of ~200) and becomes virtually independent of the temperature. At high temperatures, small values of the thermopower are associated with the ambipolarity of conduction, when the concentrations of holes and electrons involved in conduction are of almost the same order of magnitude. The absence of sign inversion of the thermopower indicates that the concentration of holes in $TlNiS_2$ always exceeds concentration of electrons involved in conduction.

The electric and thermoelectric properties of $TlMnSe_2$ and $TlMnS_2$ were studied. In the temperature range 130÷315K temperature dependence of the conductivity (σ) of $TlMnS_2$ increases exponentially with increasing temperature, i.e. σ(T) -dependence had the semiconductor nature. It was shown that σ(T) -dependence of $TlMnS_2$ consists of regions with following activation energies: 0.178 and 0.44 eV (Fig. 3).

The temperature dependence of the conductivity of $TlMnSe_2$ had the metallic nature (Fig. 4). Temperature dependence of the thermoelectromotive force (α) in $TlMnSe_2$ was studied. The thermo-e.m.f. sign corresponded to the p-type conductivity of $TlMnSe_2$ in the temperature range 88÷300K (Fig. 4, curve 2). With increasing of temperature from 88 to 300K the value of thermo-e.m.f. in $TlMnSe_2$ was increased from 77 to 200 μ V/K. At T=194 K the anomaly was revealed on the dependence α (T).

Low- temperature branch of α(T) - dependence in $TlMnSe_2$ had linear character with extapolation to T=0 according to metallic formula for thermo-e.m.f. (2).

Temperature dependences of the conductivity and thermo-e.m.f. of $TlCoS_2$ have been investigated in wide range of temperatures (Fig. 5 and 6). It was established, that $TlCoS_2$ characterized by p-type of conductivity in 77÷225K temperature interval and at 225K inversion of thermo -e.m.f. sign takes place. It was shown, that $TlCoS_2$ is ferromagnetic compound, and $TlCoSe_2$ is ferrimagnetic.

Investigation of the temperature dependence of electrical conductivity and Hall coefficient of $TlFeTe_2$ shows a band gap in $TlFeTe_2$ of a size of 0.42 eV. It was shown that scattering of current carriers on acoustic vibrations of lattice takes place at high temperatures ($\mu \sim T^{-3/2}$). Temperature behavior of thermopower. in $TlFeTe_2$ is studied. The concentration ($n_p=6.67\cdot 10^{17}$ cm^{-3}) and effective mass of hole ($m_p=0.074m_0$) are calculated for $TlFeTe_2$.

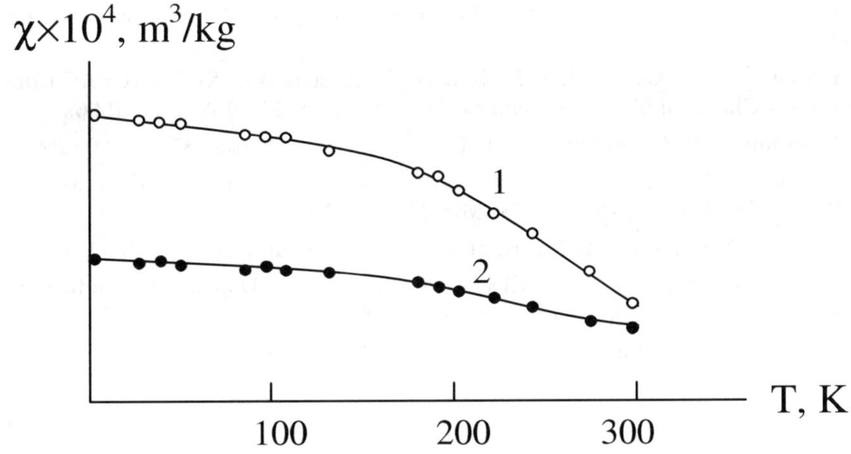

Figure 8. Dependence of magnetic susceptibility of $TlFeSe_2$ on magnetic field at T=4.2K: 1 - H||c-axis; 2 - H⊥c-axis

By study of the temperature dependence of electrical conductivity of $TlFeSe_2$, the width of the band gap in $TlFeSe_2$ was established to be 0.68 eV.

Magnetic susceptibility of $TlFeSe_2$ single crystals was investigated within 4.2-295 K temperature range.

Fig. 7 shows temperature dependence of magnetic susceptibility of $TlFeSe_2$ single crystal (χ) at magnetic field H=636kA/m when H⊥c - axis of crystal (curve 1) and $H||$c (curve 2). From these curves difference in χ values at H⊥c and $H||c$ and change of temperature behavior of χ are observed.

Dependence of χ on H at T=4.2 K for $TlFeSe_2$ is illustrated by Fig. 8. It is seen from Fig. 8 that with increasing of H, values of χ are decreased.

Obtained regularities of temperature and field dependences of magnetic susceptibility of $TlFeSe_2$ single crystal show that the magnetic properties of this crystal are common for antiferromagnetics [11].

References

[1] S.N. Mustafaeva, E.M. Kerimova and A.I. Dzhabbarly, Fiz. Tverd. Tela, **42** (12), 2132 (2000).

[2] E.M. Kerimova, F.M. Seidov, S.N. Mustafaeva and S.S. Abdinbekov. Neorg. Mater. **35** (2), 157 (1999).

[3] F.M. Seidov, E.M. Kerimova, S.N. Mustafaeva, et al., Fizika, **6** (1), 47 (2000)

[4] E.M.Kerimova, S.N.Mustafaeva, F.M.Seidov. Abstracts of 6th International School- Conference " Phase diagrams in Material Science ". Kiev. Ukraine. 14th -20th October. 2001.

[5] M.A.Aljanov, E.M.Kerimova, M.D.Nadjafzade. Fiz. Tverd. Tela, **32**(8), 1449 (1990).

[6] M.A.Aljanov, E.M.Kerimova, M.D.Nadjafzade. Abstracts of XIV th IUPAC Conference on Chemical Thermodynamics. Osaka. Japan. 25-30 August. 1996.

[7] E.M. Kerimova, R.Z. Sadikhov and R.K. Veliev. Neorg. Mater. **37**(2), 180 (2001).

[8] N.F. Mott and E.A. Davis Electronic Processes in Non-crystalline Materials.(Clarendon Press, Oxford, 1971; Mir, Moscow 1974).

[9] V.Augelli, C.Manfredotti, R.Murri, et al. Nuovo Cimento, **38** (2), 327 (1977).

[10] B.I. Shklovskii and A.A. Efros, Electronic Properties of Doped Semiconductors, Nauka, Moscow, 1979; Springer, New York, 1984.

[11] N. D. Mermin, H.Wagner. Phys. Rev. Lett. **17**, 1133-1136 (1996).

LOCALIZED MAGNETIC POLARITONS IN THE MAGNETIC SUPERLATTICE WITH MAGNETIC IMPURITY

R.T. Tagiyeva

Department of Physics, Faculty of Sciences, Ankara University, 06100 Tandogan, Ankara, Turkey and
Institute of Physics Azerbaijan National Academy of Sciences, Baku-370143, Azerbaijan

Abstract The magnetic polaritons localized at the magnetic impurity layer in magnetic superlattice composed of the alternating ferromagnetic and nonmagnetic layers are investigated in the framework of the electromagnetic wave theory in the Voigt geometry. The general dispersion relation for localized magnetic polaritons is derived in the long-wavelength limit.The dispersion curves and frequency region of the exsistence of the localized magnetic polaritons for different parameters of the superlattices and magnetic impurities are calculated numerically and analysed.

Keywords: Magnetic polariton, magnetic superlattice, magnetic impurity.

The propertices of the magnetic polaritons in the magnetic multilayer systems consisting of two and more alternating magnetic or magnetic /nonmagnetic components have attracted considerable attention during the past two decades. Magnetic polaritons, coupled electromagnetic and spin wave modes, although discussed by many authors in different ferro and antiferromagnetic arrangements, are a topic of continuing interest. Bulk and surface spin waves and magnetic polaritons in magnetic films and superlattices have been studied in the literature [1], [2], [3],[4][5], [6],[7],[9],[10],[11]. Magnetic polaritons propagating in finite ferromagnetic/non-magnetic superlattice were considered in [8].The magnetic polariton modes in metamagnet thin film and in the antiferromagnetic films, whose thickness is much larger than the interatomic distances are investigated in [12],[13],[14]. Spectrum of the magnetic polaritons localized at the junction between magnetic superlattice and magnetic material were discussed recently in [15], [16]. The aim of this

S. Halilov (ed.), Physics of Spin in Solids: Materials, Methods and Applications, 207–216.

paper to extend our previous works [17], [18] on subject, considering the propagation of the localized magnetic polaritons in ferromagnetic/non-magnetic superlattice with impurity layer. In such systems one finds both surface polaritons, in which excitation is localized near the surface, and guided modes, where excitation has a standing- wave-like character. In this case the impurity region works as waveguide, because of the magnetic polaritons propagate freely over the defect layer and damp in the perpendicular direction on either side of this region.

We consider a geometry in which the film ineterfaces are perpendicular to the x- axis, whereas the magnetization $\vec{M}_0$ and the external magnetic field $\vec{H}_0$ are applied in the z- direction. The surface polaritons propagate along the y-axis, parallel to the surface of the impurity layer (Voigt geometry) and perpendicular to the magnetic moments and to the applied external magnetic field. The impurity layer occupy the region 0¡x¡d, d being its thickness.The superlattice consists of alternating ferromagnetic films of thickness $\mathrm{d_1}$and nonmagnetic films of thickness $\mathrm{d_2}$. The elementary unit of SL have length $\mathrm{L=d_1+d_2}$. Here we neglect the dielectric properties of the magnetic material and ignore the exchange interaction.

We begin our discussion with the determination of the dynamic response of the classic ferromagnet. In the long-wavelength limit non-vanishing components of the frequency-dependent magnetic permeability tensors is given by

$$\mu_{xx}(\omega) = \mu_{yy}(\omega) = \mu_{\perp}(\omega) = \mu(1 + \frac{\Omega_0\Omega_{\rm m}}{\Omega_0^2 - \omega^2}), \tag{1}$$

$$\mu_{xy}(\omega) = -\mu_{yx}(\omega) = i\mu_x(\omega) = i\mu\frac{\omega\Omega_{\rm m}}{\Omega_0^2 - \omega^2}. \tag{2}$$

Here

$$\Omega_0 = \gamma H_0,\, \Omega_{\rm m} = \gamma 4\pi \mathrm{M}_0, \tag{3}$$

where γ is the gyromagnetic ratio and μ is the high frequency ($\omega \gg \Omega_0$) permeability, caused by the magnetic dipolar excitations other than the spin wave excitations (i.e. optical magnons).

In order to find the polariton spectra, we use the Maxwell's curl equations in the magnetic film. After eliminating the electrical field variable $\vec{E}$, we obtain the following wave equation:

$$\nabla^2\vec{h} - \vec{\nabla}(\vec{\nabla}\vec{h}) - \frac{1}{c^2}\frac{\partial^2}{\partial t^2}(\,\vec{h} + 4\pi\vec{m}) = 0, \tag{4}$$

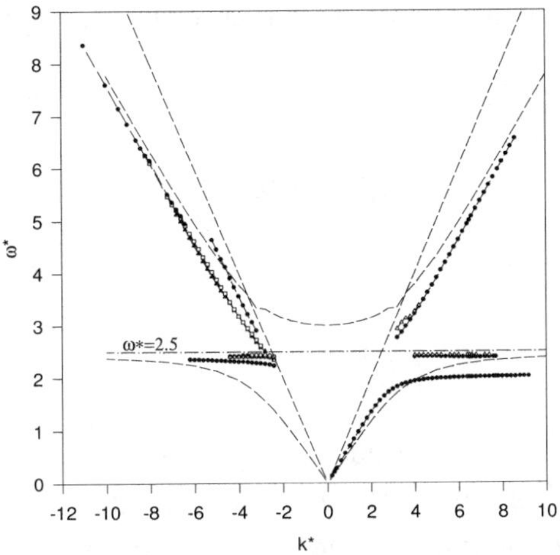

Figure 1. The dispersion curves of the surface/guided modes of the magnetic polaritons for SL (Ni/non) with impurity Fe. Here $\omega* = \frac{\omega}{\Omega_m^{Fei}}$ and $k* = \frac{ck}{\Omega_m^{Fei}}$ .We use the parameters : $d_1^* = d_2^* = d_3^* = d_4^* = 0.1$ and $d_0^* = 0.5$.The broken lines denote the dispersion curves of bulk polaritons in Fe and photon lines.

where c is the light velocity in the vacuum.

We consider only the transverse electric (TE) mode in which $\vec{E}$ has a nontrivial component only in the z direction. Here $\vec{h}$ and $\vec{m}$ are the dynamical components of the magnetic field and the magnetization, respectively and can be written in the form $\vec{h},\vec{m} \sim \exp(i\vec{k}_{\|}\vec{r}_{\|})$, where $k_{\|}^2 = k_y^2 + k_z^2$- two-dimensional wave vector (our system is translationally invariant in the y and z directions). In the Voigt configuration $k_z=0$ and $k_{\|} = k_y = k$.

The general solution of the equation 4 can be written in the form:

$$h_i(r,t) = (B_i^+(n)e^{\beta x'} + B_i^-(n)e^{-\beta x'})e^{i(ky-\omega t)}; i = x, y; \tag{5}$$

x'=x-nL for magnetic film of SL ($L = d_1 + d_2$) and

$$h_i(r,t) = (A_i^+(n)e^{\alpha x'} + A_i^-(n)e^{-\alpha x'})e^{i(ky-\omega t)}; i = x, y; \tag{6}$$

for the non-magnetic film of SL and in the impurity layer

$$h_i^{(0)}(r,t) = (C_i^+ e^{\beta_0 x} + C_i^- e^{-\beta_0 x})e^{i(ky-\omega t)}; i = x, y. \tag{7}$$

The non-trivial solution of Eq. 4 and the divergence condition div $\vec{b} = 0$ leads to

$$\alpha^2 = k^2 - \frac{\omega^2}{c^2}, \tag{8}$$

$$\beta_j^2 = k^2 - \frac{\omega^2}{c^2}\mu_v^{(j)}, \tag{9}$$

The parameter β_0 is imaginary for guided modes and real for surface modes . Here $\mu_v^{(j)}$is called the effective magnetic permeability of the j–th ferromagnetic medium in the Voigt geometry (we assume $\mu_v^{(j)} = 1$ for non-magnetic material) and is defined by

$$\mu_v^{(j)} = \mu_\perp^{(j)} - \frac{\mu_x^{(j)2}}{\mu_\perp^{(j)}}, \tag{10}$$

From the divergence condition div $\vec{b}$=0, we can derive the relation for the constants $B^{\pm}_{x(y)}$ (n) , $A^{\pm}_{x(y)}$(n) and $C^{\pm}_{x(y)}$:

$$\frac{B_x^{\varepsilon}(n)}{B_y^{\varepsilon}(n)} = i\frac{\frac{\omega^2}{c^2}\mu_x(\omega) - \varepsilon k\beta}{k^2 - \frac{\omega^2}{c^2}\mu_\perp(\omega)}, \tag{11}$$

$$\frac{A_x^{\varepsilon}(n)}{A_y^{\varepsilon}(n)} = -i\frac{\varepsilon k}{\alpha}, \tag{12}$$

$$\frac{C_x^{\varepsilon}}{C_y^{\varepsilon}} = i\frac{\frac{\omega^2}{c^2}\mu_x^{(0)}(\omega) - \varepsilon k\beta_0}{k^2 - \frac{\omega^2}{c^2}\mu_\perp^{(0)}(\omega)}, \tag{13}$$

here $\varepsilon = \pm$.

Applyiing the boundary continuity condition for tangential component $\vec{h}$ and the normal component $\vec{b} = \vec{h} + 4\pi\vec{m}$ to the left (x=0) and right (x=d) boundaries of the impurity layer, we obtain the following relations between the amplitudes:

$$\widetilde{B_y^+}\widetilde{B_y^-} = \mathrm{R_2 C_y^+ C_y^-} = \mathrm{R_2 R_1 B_y^+ B_y^-} = \mathrm{T^s B_y^+ B_y^-}, \tag{14}$$

where $T^s = R_2R_1$ is the transfer matrix across the impurity layer and given by the following expression:

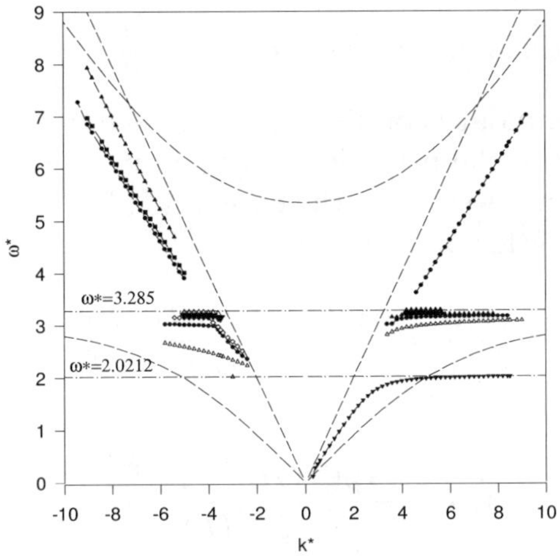

Figure 2. Same as in fig.1, but now for SL(Ni/non) with impurity Gd.

$$T^s_{11(22)} = A\exp(\mp\beta \mathrm{d})\left[\begin{array}{c} 2\mu_\perp^{(0)}\beta_0\mu_\perp\beta\cosh(\beta_0 \mathrm{d}) \pm \sinh(\beta_0 \mathrm{d})(\mu_\perp^{(0)}\mu_\mathrm{v}^{(0)}\gamma + \\ 2\mu_x^{(0)}\mu_x k^2 - \frac{\gamma_0}{\gamma}((\mu_x k)^2 - (\mu_\perp\beta)^2)) \end{array}\right], \tag{15}$$

$$\mathrm{T}^\mathrm{s}_{12(21)} = \pm \mathrm{A}\exp(\mp\beta \mathrm{d})\sinh(\beta_0 \mathrm{d})\left[\begin{array}{c} \mu_\perp^{(0)}\mu_v^{(0)}\gamma + 2\mu_x^{(0)}k(\mu_x k \pm \mu_\perp\beta) \\ -\frac{\gamma_0}{\gamma}(\mu_x k \pm \mu_\perp\beta)^2 \end{array}\right], \tag{16}$$

where $\mathrm{A}=\frac{1}{2\mu_\perp\beta\mu_\perp^{(0)}\beta_0}$ and the parameters were defined as:

$$\gamma_j = k^2 - \frac{\omega^2}{c^2}\mu_\perp^{(j)}. \tag{17}$$

The presence of the impurity layer leads to appearance of localized magnetic polaritons. In order to have a bounded excitation , we require that the transverse component of the wavevector is imaginary or complex for superlattice. Only in this case we may find surface magnetic polaritons in the frequency gaps between the bulk bands.

From the condition of solvability of the equation 14 we can obtain the following expression:

$$W_1(T^s_{11} + T^s_{12}W_2) - T^s_{22}W_2 - T^s_{21} = 0, \tag{18}$$

where the functions $W_1 = \frac{\widetilde{B_y^-}}{\widetilde{B_y^+}}$ and $W_2 = \frac{B_y^-}{B_y^+}$.

Applying the boundary conditions for $\vec{h}$ and $\vec{b}$ to the interface berween the layers 1 and 2 of the n-th elementary unit of the SL and by using the Bloch's theorem , we derive the expression for the functions W_1 and W_2 in the following form:

$$W_1 = \frac{\exp(-QL) - T_{11}(L)}{T_{12}(L)}, \tag{19}$$

$$W_2 = \frac{\exp(-QL) - T_{11}^{-1}(L)}{T_{12}^{-1}(L)}, \tag{20}$$

where matrix T(L) is the transfer matrix with det T=1 and T^{-1} is the inverse of matrix T. The expressions for the elements of the transfer matrix are given in Appendix. The parameter Q is the decay parameter of the magnetic polaritons in the superlattice describing the waves damping along the axis of the SL and $(\text{ReQ})^{-1} > 0$ is the penetration depth of the magnetic polaritons into the superlattice. One can define the decay parameter from the dispersion equation for the damping magnetic polaritons in ideal superlattice [1],[2],[3]:

$$\begin{aligned}\cosh(QL) &= \cosh(\beta d_1)\cosh(\alpha d_2)\\ &+ \left[\frac{\gamma}{2\alpha\beta\mu_\perp(\omega)} + \frac{\alpha\mu_v(\omega)}{2\beta}\right]\sinh(\beta d_1)\sinh(\alpha d_2)\end{aligned} \tag{21}$$

The equation 18 is the main dispersion relation for localized magnetic polaritons. Solving the equation 18 together with 19 and 20 one can find the relation between the frequencies of the localized magnetic polaritons and the wavevector. Only those solutions of Eq. 18 for which the condition Q> 0 are fulfilled describe physical localized magnetic polaritons. In the magnetostatic limit $c \to \infty$, the equation 18 reduces to the dispersion equation for localized magnetostatic waves.

We now discuss our results for specific cases of the magnetic superlattice SL (Ni/ nonmagnet) with impurity layer Fe($\mu_0 M_0^{Fe} = 2.16T$, g =2.15; here g denotes the Lande factor and μ_0 is the magnetic permeability of the vacuum) and Gd ($\mu_0 M_0^{Gd} = 7.12\ T, g^{Gd} = 2.17$), respectively.

For numerical calculation we have introduced the following dimensionless parameters:

$\Omega_0^* = \frac{\Omega_0}{\Omega_m^{Fe}}$; $\omega^* = \frac{\omega}{\Omega_m^{Fe}}$; $k^* = \frac{ck}{\Omega_m^{Fe}}$; $d^* = \frac{\Omega_m^{Fe} d}{c}$;

We use the following parameters: $\Omega_0^* = 2.$, $\mu = 1.75$, $d_1^* = d_2^* = d_3^* = d_4^* = 0.1$ and for Ni ($\mu_0 M_0^{Ni} = 0.6084\ T$, $g^{Ni} = 2.187$) Fig.1 shows the magnetic polariton dispersion curves of the bulk and surface/guided modes for magnetic impurity with Fe for two directions of the wavevector. Here the bulk and surface/guided modes frequencies $\omega^* = \frac{\omega}{\Omega_m^{Fe}}$ are shown as a function of the reduce wavevector $k^* = \frac{ck}{\Omega_m^{Fe}}$. The broken lines denote the dispersion curves of bulk polaritons in ferromagnet system (in this case Fe) and photon lines. As it is seen, there are an infinite number of spectral branches $\omega = \omega_n(k)$. In the frequency range $\omega < \omega_\perp^{(Fe)} = \left[\Omega_0^{(Fe)}(\Omega_0^{(Fe)} + \Omega_m^{(Fe)})\right]^{\frac{1}{2}}$ ($\omega_\perp^{*(Fe)} = 2.45$) there are two types of the magnetic polariton branches. The lower branch of the localized magnetic polaritons degenerates at the small value k^* with the group velocity $V_g = \frac{d\omega}{dk} \to c$, while for large value of the wavevector the group velocity tends to zero.The lower branch has an asymptotic frequency $\omega_{\text{lim}}^* = 2.0212$. There is a gap between the lower and upper branches, which tends to zero with increasing number n. The group velocity of the upper branches approaches to zero for all values of the wavevector. The frequencies of the lower and upper branches lie in the region where the parameter $\beta_0(k)$ is imaginary and $\alpha(k)$ is real and positive. These modes result from the guided modes of separate impurity layer. With increasing value k^* the lower branch $\omega = \omega_1(k)$ passes to the range where the parameter $\beta_0(k)$ and $\alpha(k)$ both real and positive and the wave amplitude varies exponentially when one moves from the interface into the impurity layer. Therefore, with increasing k^* the guided mode fransforms to the surface mode. For $k \to \infty$ the upper branches come into the lower branch curve of the bulk mode of the polaritons in an infinite Fe $\omega^* = \frac{k^*}{\sqrt{\mu_v}}$.

As one can see, in the frequency regine $\omega > \omega_s^{Fe}$, where $\omega_s^{(i)} = \Omega_0^{(i)} + \frac{\Omega_m^{(i)}}{2}$ is the Damon-Eshbach wave frequency ($\omega_s^{*Fe} = 2.5$) we have SM branches which start at the photon line at $\omega^* = 2.761$ and then come into the upper branch curve of the bulk $\mid e \mid$ of the polaritons.

For k¡0 it is posible to distinguish two different types of the localized modes: those, which lie under the limit frequency $\omega_\perp^{(Fe)}$ and merge into the lower bulk branch as $k^* \to -\infty$ (guided modes) and other one(high-frequency SM) , which start at the finite value of the frequency $\omega = \omega_c > \omega_s^{Fe}$ on the photon line and then merge into a upper bulk region with increasing $|k^*|$. As it is seen, the spectrum of the localized magnetic polaritons is non-reciprocal with recpect to propagation direction, i.e. $\omega(-k) \neq \omega(k)$.

For the case where the impurity layer is Gd, the dispersion curves of localized magnetic polaritons are shown in fig.2. The behavior of the surface/guided modes is simular to those found for previous case. It is essensial that the varying of the impurity material does not change the frequency range of the lower branch of the localized polaritons ($\omega^*_{\rm lim}$ = 2.0212), whereas the frequency region of the upper guided modes branches is determined by the physical parameters of the impurity layer. From the numerical calculations, it is obtained that the asymptotic frequency for upper branches increases with increasing the spontaneous magnetization of the impurity material.

For completeness, we have also shown in fig. 3 the allowed magnetic polaritons modes, propagating in the SL(Gd/non) with impurity layer Fe. Compared with fig.1 , it is easy to see that for k¡0 direction, the lower branch, which lies under the lower branch of the bulk modes in infinite ferromagnet (Fe), split into two parts, betwen them lies a forbidden wavevector gap. These pure surface modes appear in the restricted frequency range. Note, that here the guided modes appear also in the high-frequency region between photon line and upper bulk branch. Along the k¿0 direction we also see a several curves, starting at $k^* = 0$. Now the lower branch has an asymptotic frequency $\omega^* = 1.9027$, while the upper high-frequency branch starts at $\omega^* = 3.9519$. Here we see a few pure gueded modes curves and one SM branch which merge into the lower branch curve of the bulk mode of the polaritons in an infinite Fe as k$\rightarrow +\infty$.The non-reciprocal nature of the localized polaritons is clearly evident.

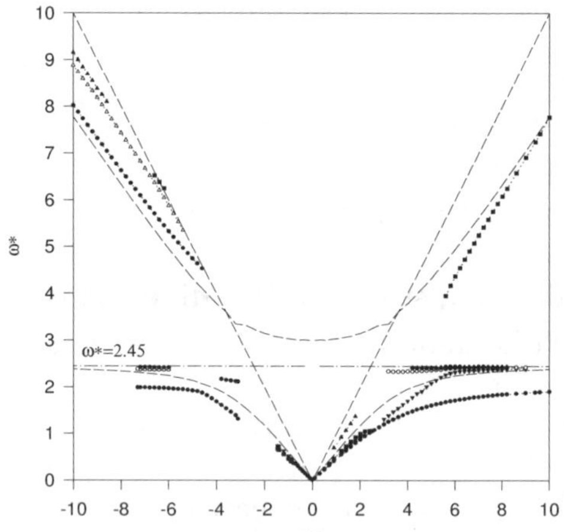

Figure 3. The dispersion curves of the surface-guided modes of the magnetic polaritons for SL (Gd/non) with impurity Fe .

Thus, the varying of the magnetic component of SL leads to the varying of the limiting frequency of the lower branch and starting point of high-frequency branch of the surface magnetic polaritons.The frequency region of the exsistence of the guided modes branches of the localized magnetic polaritons can be changed by varying of the material of the impurity layer. From the numerical calculations, it is obtained that the asymptotic frequency for guided upper branches increases with increasing the spontaneous magnetization of the impurity layer.

The most appropriate experimental technique to probe the magnetic polariton modes is the inelastic light scattering spectroscopy of Raman and Brillouin type. We hope that our theoretical prediction can be tested throughout these experimental measurements.

Acknowledgments

I thank Dr. I.N.Askerzade for useful discussions and help.

Appendix

The elements of the transfer matrix appearing in equation 19 are

$$T_{11(22)} = \frac{1}{2}e^{\mp\beta d_1}\left\{2ch(\alpha d_2) \pm sh(\alpha d_2)\left[\frac{\gamma}{\alpha\beta\mu_\perp} + \frac{\beta\mu_v}{\beta}\right]\right\}. \qquad (22)$$

$$T_{12(21)} = \frac{1}{2}e^{\mp\beta d_1} sh(\alpha d_2)\left\{\pm\frac{\gamma}{\alpha\beta\mu_\perp} \mp \alpha\frac{(\mu_\perp\beta \pm \mu_x k)^2}{\beta\mu_\perp\gamma}\right\}. \qquad (23)$$

References

[1] J. Barnas, Solid State Commun. **61** (1987) 405.

[2] J. Barnas, J. Phys. C.: Solid State Phys. **21** (1988) 4097

[3] J. Barnas, Condens. Matter **2** (1990) 7173.

[4] A. Karsono & D. Tilley, J. Phys. C **11** (1987) 3487.

[5] C. Zhou & C. Gong , Phys. Rev. B **39** (1989) 2603.

[6] N. Raj & D. Tilley, Phys. Rev. B **36** (1987) 7003.

[7] N. Raj & D. Tilley, Phys. Status Solidi b **152** (1989) 135

[8] C. Shu, Z. Peng & Z. Ming, Phys. Letters A **12** (1993) 195.

[9] R. Krishnan, A. Das, N. Keller, H. Lassri, M. Porte, M. Tessier, J. Magn, and Magnetic Mat. **174** (1997) L17.

[10] R. E. Camley, J. Magn, and Magnetic Mat. **200** (1999) 583.

[11] Ivan K. Schuller, S. Kim, C. Leighton, J. Magn, and Magnetic Mat. **200** (1999) 571.

[12] E. S. Guimaraes, E. L. Albuquerque, Physica A **277** (2000) 405

[13] E. S. Guimaraes, E. L. Albuquerque, Solid State Communication **122** (2002) 623

[14] E. S. Guimaraes, E. L. Albuquerque, Phys. Letters A **307** (2003) 172.

[15] R. Tagiyeva and M. Saglam, Solid State Communication **122** (2002) 413.

[16] R. Tagiyeva and M. Saglam, Physica E **16** (2003) 355.

[17] R. Tagiyeva , Y. Seidov & N. Hashimzade, J. Magn, and Magnetic Mat. **136** (1994) 88.

[18] R. Tagiyeva , Magnetic and superconducting materials (MSM) vol. B (2001) 889 Ed. Kitazawa, Akhavan, Jensen, Singapour.

SPIN STABILITY AND LOW-LYING EXCITATIONS IN SR_2RUO_4

S. V. Halilov[1,2], D. J. Singh[1], A. Y. Perlov[3]

[1] *Center for Computational Materials Science, Naval Research Laboratory, Washington, DC 20375*

[2] *Department of Materials Science and Engineering, University of Pennsylvania, Philadelphia, PA 19104*

[3] *University of Munich, Germany*

Abstract LSDA non-relativistic frozen spin wave calculations reveal the presence of a non-uniform spin configuration in the ground state of Sr_2RuO_4, characterized by a non-vanishing local spin moment on Ru hosts. This is related to nesting of the essentially two-dimensional Fermi surface. Spin instability at nesting vectors is manifested as a spin-density wave in the LSDA. By including the spin-orbit coupling, the magneto-crystalline anisotropy is estimated to be vanishingly small in spite of the layered structure of the system. On account of the small anisotropy, it is speculated that the zero-point fluctuations can destroy the staggered magnetization in this highly two-dimensional system. The static paramagnetic susceptibility is found to be slightly anisotropic dominating within tetragonal basal plane. The paramagnons intrinsically featured with phasons can tentatively be considered as a vague assessment for the lower part of excitation spectrum. As possible mediators for the superconductivity, the phasons reflect a specific symmetry of the ground state spinor: on an atomic scale the spinor symmetry is triplet, but on the scale of the helix the symmetry is a singlet. The obtained LDA results show the pure system is on the borderline between paramagnetism and non-uniform incommensurate antiferromagnetic instability.

Keywords: Spin-density wave, quantum critical fluctuations, spin-orbit coupling, magnetic anisotropy

15.1 Introduction

The unconventional superconductor Sr_2RuO_4 is known to have strong incommensurate antiferromagnetic (AFM) fluctuations at $q\sim(0.30.30)2\pi/a$

S. Halilov (ed.), Physics of Spin in Solids: Materials, Methods and Applications, 217–242.

with the same intensity below and above the critical (superconducting) point T $\sim$ 1.5 K [1]. The spin-spin correlation length is of order of 10 Åin the ab-plane, with no correlation between planes, but little is known about the quasi-particles which mediate the formation of the superconducting order parameter below $T_C \sim 1.4$ K. Several attempts have been made to find out whether the lattice vibrations or/and magnetic fluctuations might be involved in the phenomenon. Replacement of Sr by essentially smaller cation Ca leads to a transition to a so-called tilted phase, which is characterized by the Neel AFM ordering and is a Mott insulator as the end member Ca_2RuO_4. This is because of orthorhombic distortions caused by the zone-boundary rotational mode, favored by the larger space made available for rotations by the small Ca ion. Superconductivity, however, doesn't survive even a small concentration of Ca, and this is apparently related to the impurity scattering deteriorating the coherent state of charge transportation. Inelastic neutron scattering [2], shows the Σ_3 phonon branch, corresponding to the RuO_6 octahedron rotation around c, exhibits a drop near the (0.5 0.5 0) zone boundary. That is, Sr_2RuO_4 is close to a rotational instability. On the other hand, the frequency of the rotational mode depends little on temperature, i.e. there is no sign of mode softening. Flat dispersion along the (0.5 0.5 ξ) zone boundary shows that there is almost no coupling between rotational deformations of neighboring layers, thus emphasizing the 2D-character of the mode. The other rotational mode Σ_4 which may be viewed as a tilting around an axis in *ab*-plane, does not exhibit any anomalous behavior. The Σ_3 mode gets considerably stiffer with cooling, which cannot fully be explained by thermal contraction. Neutron-powder diffraction [3] investigation of the thermal expansion and compressibility of Sr_2RuO_4 shows that the temperature dependence of the Ru-O2 apical bond length is linear with no structural anomaly. No structural anomaly has been found for any other structural parameter, providing no evidence for a metal-to-insulator transition within a wide temperature range. It was theoretically [4] shown earlier, that the Raman zone-center mode and rotational zone boundary Σ_3 mode are not sensitive to the hydrostatic pressure up to $P \approx 2.4GPa$, although Σ_3 undergoes a softening at a specific non-hydrostatic pressure.

The zone-boundary rotational instability manifests also on the surface: low-energy electron diffraction data from Sr_2RuO_4 indicate that there is a $\sqrt{2}$ x $\sqrt{2}$ surface reconstruction induced by the freezing of the soft zone boundary phonon into a static lattice distortion , and comparison with band structure calculation predicts that the resulting surface is ferromagnetic (FM) [5]. Although, angle-resolved photoemission [6] shows that the electronic structure of the layered perovskite Sr2RuO4 is most readily explained by the surface reconstruction, no direct evidence of the FM ordering was found.

In the bulk [1], a substantional broadening of the spin fluctuation with a rate of 10 meV$\sim$ $100 T_C$ might be used as indication of p-wave superconductivity. However, the experiment cannot readily fit any of the existing theoretical models of superconductivity: multiband theories predict a resonance of susceptibility, but the experiment shows no resonance at all. The magnetic form-factor of Ru indicates a rather delocalized or disordered magnetization density with in-plane correlations.

Recent elastic neutron scattering measurements confirmed the development of an incommensurate AFM structure upon partial replacement of Ru by Ti. Even a small concentration of Ti of $\sim$ 2 percent makes the system unstable against building a SDW of sinusoidal type in the ground state [7] with the same Fermi-surface nesting origin as the peak at the incommensurate vector in the unsubstituted compound. The spin-spin correlation length of the excitations determined from the half-width of spin susceptibility, is comparable to that of the pure compound but is much shorter than the correlation of the elastic order. The latter number is about 50 $\mathring{A}$ and was claimed to correlate with the respective concentration of Ti, which also corresponds to the average distance between neighbour sites of Ti in the alloy. There is no sign of lattice reconstruction, which makes the system very attractive for investigation of the trends associated with quantum critical point behavior of the spin degree of freedom. Altogether, pure strontinum ruthenate seems to have a strong propensity toward developing a non-uniform magnetic structure in the ground state on the borderline between a paramagnetic and incommensurate AFM configuration.

The present paper reports results of DFT-LSDA calculations for the ground state of bulk Sr_2RuO_4, which appears to be a SDW of sinusoidal type with nonrelativistic treatment. Section 15.2 gives the details of the non-relativistic DFT-LSDA calculations for the ground state of the pure compound. Formation conditions for spin and orbital moments are the subject for Section 15.3, where in particular the claim is made that a vanishingly small magnetocrystalline anisotropy enables zero-point fluctuations to destroy the staggered magnetizations. Section 15.4 presents relativistic results for the static magnetic susceptibility, which turns out to be anisotropic slightly favoring the tetragonal basal plane. Section 15.5 serves as a summary and speculates about a the possible impact on superconductivity of magnetic low-lying excitations of the phason nature.

15.2 LSDA magnetic ground state

As Sr_2RuO_4, the longitudinal and transverse spin degrees of freedom are coupled at equilibrium, but become decoupled under negative pressure or suppressed under positive pressure. Theoretically [8, 9] and experimentally it is well known that the Fermi surface of strontium ruthenate is strongly influenced by tetragonal crystal fields and is characterized by considerable nesting. The most prominent nesting takes place along the [110] direction at approximately [0.3,0.3,0] 2 π/a, and is built from t_{2g} states with d_{xz}/d_{yz} character centered on Ru sites and *p*-states of the planar oxygen (α- and β-sheets). The density of states (DOS) at the Fermi level derived by different band methods: LAPW, full-potential LMTO and LMTO-ASA with empty spheres, is about 4.2 st/eV/cell and large enough to cause a high spin susceptibility. Therefore, keeping in mind that the itinerant oxygen p-states act as mediators for inter- site super-exchange interaction, the presence of the incommensurate nesting suggests the formation of a spin super-structure in the ground state.

Indeed, performing the LSDA calculations with the spin density vector as a basic functional variable, we find that the variational procedure yields non-uniform spin configuration as favorable for Sr_2RuO_4 at the experimental value for the lattice parameter. Equilibrium positions of Sr and apical oxygens have been determined by earlier investigation of the lattice instability and appear to be very close to that obtained from experiment. Fig. 1 illustrates total energy per formula unit and magnetization at Ru sites as a function in the space of the ferromagnetic cones with various pitch angles and momenta along [110] direction in the first BZ.

The procedure of energy variation over the frozen spin waves of transverse type is rather straightforward within spin-density matrix formalism as long as spin-orbit coupling is not included. In this case the problem is formally solvable in the spin-restricted crystal unit cell since the spin degree is not coupled to the lattice. Yet, a search for the energy minimum in the space of frozen magnons is only relevant if the system can be described by the Heisenberg Hamiltonian, a condition that is hard to verify in an itinerant system prior to calculations. The motivation for the energy variation over the frozen magnons is a statement made by Lyons and Kaplan in 1960 [10], that the ground state of any Heisenberg magnetic system with equivalent atoms in the basis is a simple spiral, given no anisotropy is present; an ordinary FM or the Neel AFM configuration are just a particular cases. In a comprehensive theory, this simple picture can be distorted due to anisotropy forces, which are usually of the relativistic nature. To avoid an ambiguity in reading, we will

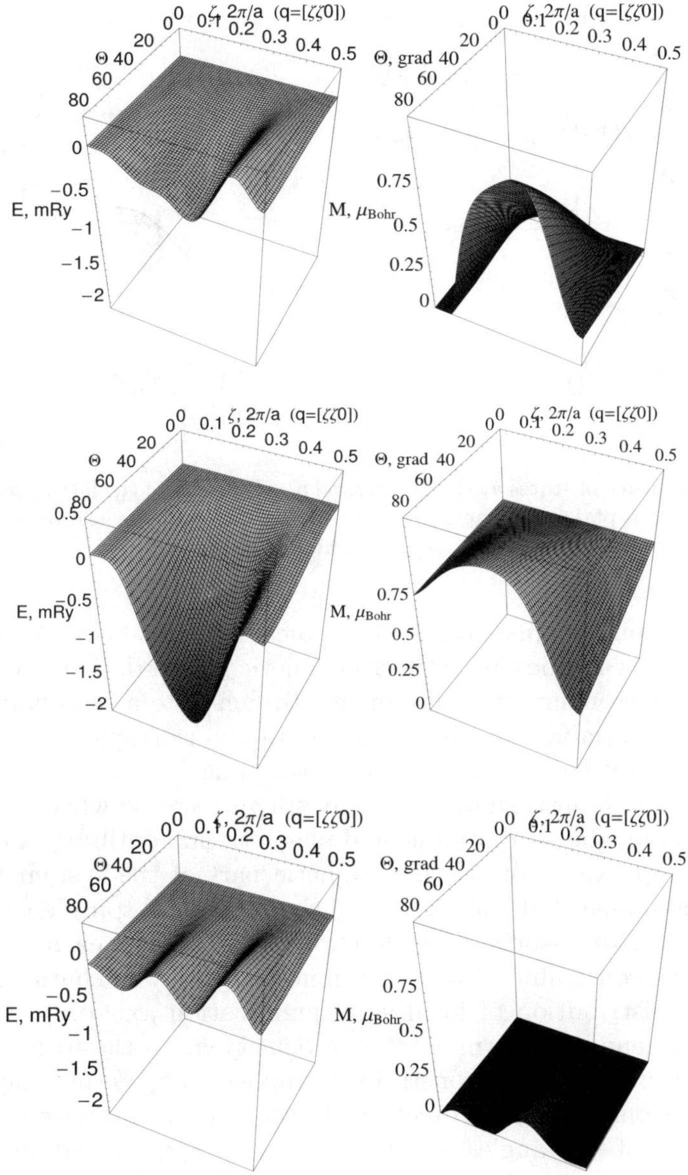

Figure 1. Total energy and magnetization M_{Ru} on Ru-site as a function of spin cone wave vector q along $[0.5,0.5,0]2\pi/a$ and pitch angle θ, at equilibrium, extended and reduced lattice parameter $a = 7.29$, $a = 7.48$, and $a = 6.92$ Bohr, respectively. There is an energy minimum and respective magnetization maximum at $\mathbf{q}_N \approx [0.35, 0.35, 0]2\pi/a$, when the Ru moments are in-plane ($\Theta = \pi/2$). Stabilization of the AFM helix under negative pressure and suppression of the magnetic ordering under positive pressure is obvious.

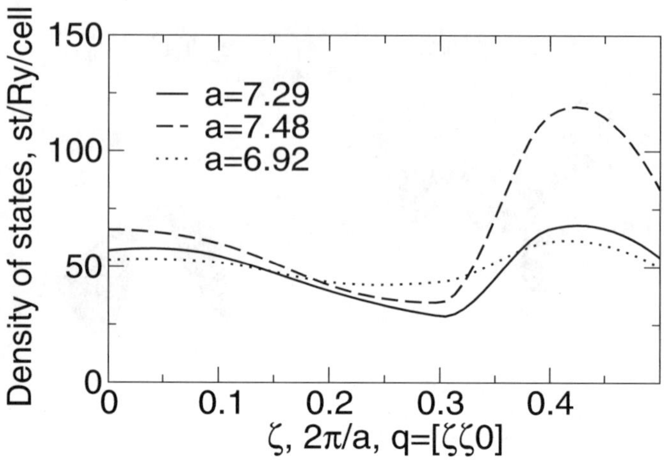

Figure 2. Density of states at the Fermi level as a function of the lattice parameter a (units of Bohr), in-plane AFM helix set. Minimum which is at the nesting, corresponds to the maximum of the Ru-site exchange splitting

label simple spiral as just spiral, and conical spiral as helix. Macroscopically the former implies an antiferromagnetic spin ordering, whereas the latter has a ferromagnetic component. To facilitate a reasonable choice of basic spin variables and investigation of the isotropic exchange part of system Hamiltonian, the spin-orbit coupling effects will first be excluded. We will later turn to the relativistic interaction when focusing on the effects of magneto-crystalline and magnetic susceptibility anisotropy. As a first approximation to the magnetic part of the system Hamiltonian, it is assumed that there is only a single integral spin-density vector $\mathbf{S}_i = \mathrm{Tr} \int_{V_i} \hat{n}(\mathbf{r}) d\mathbf{r}$ associated with a volume V_i around each atomic site i, so that the constrained frozen magnon energy is well characterized by the spatial distribution of local spin quantization axis $\mathbf{e}_i = \mathbf{S}_i / S_i$ and respective eigenvector norm S_i of spin density $\hat{n}$. In the frozen magnon approximation, a periodic constraint is applied to $\mathbf{e}_i$ so that the internal energy E becomes a unique function. In the polar coordinate frame, this is a function of two angles given as $\theta_i = \arccos \mathbf{z}_i \cdot \mathbf{q}$ and $\phi_i = \mathbf{q} \cdot \mathbf{r}_i$, with $\mathbf{z}$ normal to the plane of spin quantization axis rotation and $\mathbf{q}$ a static magnon vector. Note that the choice of $\mathbf{z}$ is arbitrary as long as no spin-lattice coupling is included. If anisotropy effects are included, the aligning $\mathbf{z}$ along the anisotropy axis would be natural.

Fig. 1 clearly demonstrates the presence of minima on the energy surface $E(\theta, \mathbf{q})$ at $\mathbf{q} \approx \mathbf{q}_N$ and $\theta = 90^0$. The magnetization $M_i = g\mu_{\mathrm{Bohr}} S_i$ shows a maximum for the same spin configuration parameters. Therefore, the anticipated frozen magnon with $\mathbf{q} \approx \mathbf{q}_N$ is in fact favored by the system. Fig. 1 shows also that the spin ordering is highly sensitive to pressure: the non-uniform spin structure is stabilized by lattice extention, and conversely the local spin polarization is destroyed by uniform contraction of the system. The energy scale of the frozen magnon grows deeper from -1.0 mRy at equilibrium to -1.6 mRy with a lattice parameter increase of 3 percent using the paramagnetic energy as a reference point, and the magnetic moment at the Ru sites is increasing from 0.8 μ_{Bohr} to 1.2 μ_{Bohr}. As is obvious from Fig. 2, DOS at the Fermi level also reaches its minimum at the same spiral parameters, due to local polarization which splits the electronic states and lowers the total DOS. In fact, this minimum is even more pronounced in case of a static SDW caused by antiferromagnetic ordering, which should be considered a strong indication that the system will undergo a metal-insulator phase transition.

To estimate the size of cluster that would be sufficient to reproduce the nesting featured exchange interaction, the LSDA-derived energy of the helix $\theta, \mathbf{q}$ depicted in Fig. 1 was mapped onto a general form bilinear in terms of spins. That is, a frozen spin wave energy for a Heisenberg system with a single spin variable per site

$$\mathbf{M}_i = M\{\sin\theta\cos\mathbf{q}\mathbf{r}_i, \sin\theta\sin\mathbf{q}\mathbf{r}_i, \cos\theta\}$$

has to be

$$\begin{aligned} E^{\mathrm{spiral}}(\theta,\mathbf{q}) &= -\sum_{ij}{}' J_{ij}M^2\sin^2\theta\cos\mathbf{q}(\mathbf{r}_i-\mathbf{r}_j) \\ &= E_0 + 2\sum_{ij} J_{ij}M^2\sin^2\theta\sin^2\frac{1}{2}\mathbf{q}(\mathbf{r}_i-\mathbf{r}_j), \end{aligned} \tag{1}$$

which implies that average spin momentum squared $\langle\hat{S}_i^2\rangle$ should be an constant of the motion. However, $\langle\hat{S}_i^2\rangle$ in our system apparently is not a constant since the magnetization has considerable variation in the space of helices $\theta, \mathbf{q}$. A numerical fit in the form of the Fourier expansion over the modulation parameter q

$$M(\theta,\mathbf{q}) = \sin\theta\sum_i(\alpha_i\sin(\mathbf{q}\mathbf{r}_i+\beta_i),$$

is able to quite accuratly reproduce the magnetization in the phase space, Fig. 1. A reasonable mapping for the helix energy surface in the same space has been made in accord with expression (1) for a helix in the Heisenberg form. An LSDA-derived energy (mRy) expansion at equilibrium lattice parameter in terms of modulation vector $\mathbf{q} = q[110]2\pi/a$, where q is varying between 0 and 1/2 within the first Brillouin zone,

$$E^{\text{spiral}}(\theta, \mathbf{q}) / \sin^2\theta = E_0 + 2M^2 \sum_{ij} J_{ij} \sin^2 \pi q(n_{ij} + m_{ij})$$
$$\approx E_0 + 0.1 \sin^2 1q\pi - 1.2 \sin^2 2q\pi$$
$$+0.1 \sin^2 3q\pi + 0.6 \sin^2 4q\pi - 0.5 \sin^2 5q\pi \quad (2)$$

shows that the largest contribution comes from Ru sites located along the nesting vector. This is because the terms with even and odd numbers of πq correspond to the exchange interaction between nearest and next-nearest Ru sites along $\mathbf{q}_n \sim [110]$ chain of Ru, respectively. Obviously the long-range nesting-imposed RKKY mechanism is responsible for the spiral ordering with $J \approx -0.25$ mRy/μ^2_{Bohr} between next-nearest Ru moments, whereas the spin modulation along Ru-O-Ru chains, with much smaller $J \approx 0.02$ mRy/μ^2_{Bohr} between nearest Ru moments, is mediated by super-exchange [11]. Of course, separation of the exchange into these two mechanisms is more conventional than can be strictly justified, particularly on account of $\langle \hat{M}_i^2 \rangle$ not being a constant. On the contrary, a similar energy expansion for a stretched lattice, middle picture on the Fig. 1,

$$E^{\text{spiral}}(\theta, \mathbf{q}_n) / \sin^2\theta \approx E_0 - 1.1 \sin^2 1q\pi - 3.9 \sin^2 2q\pi$$
$$+0.7 \sin^2 3q\pi + 1.0 \sin^2 4q\pi - 0.3 \sin^2 5q\pi \quad (3)$$

shows that super-exchange, with $J \approx -0.10$ mRy/μ^2_{Bohr}, and RKKY mechanism, with $J \approx -0.33$ mRy/μ^2_{Bohr}, have different behaviors with lattice parameter change. There are two types of spin-spin correlations $\langle \mathbf{S_i S_j} \rangle$: 1) the dominant RKKY type correlations on the $\xi_{\text{RKKY}} \sim q_N^{-1} \sim 3\sqrt{2}a \sim 15$ Å distances; 2) essentially weaker ligand-type correlations on the scale of lattice parameter $\xi_{\text{ligand}} \sim a$. With lattice expansion, the Ru moments grow, a trend which hints at the impurity model and will be the subject of the next section. Therefore, the energy gain associated with RKKY interaction is increasing roughly as $M^2 N(E_F)$ with increasing a, whereas the correlations $\langle \mathbf{S_i S_j} \rangle$ remain unchanged. On the other hand, the energy gain due to the ligand exchange is on the order of $-t^2/U$, with t as the ligand hopping integral, much smaller than the former but also more sensitive to the lattice parameter. The super-exchange correlations follow the increase of M^2 but get weaker on account of diminishing t with lattice expansion, so that it becomes rather obscure to predict the resultant effect in simple terms of the perturbation. In any case, the dependence on lattice parameter changes as well as thermodynamics are dominated by RKKY type of correlations.

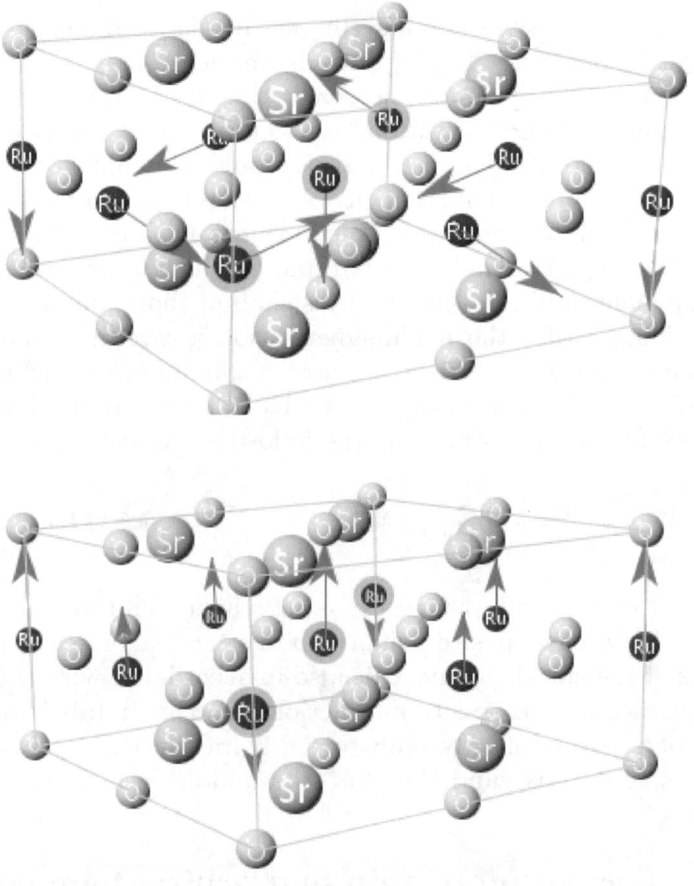

Figure 3. Magnetic moment distribution in body-centered tetragonal layered structure of Sr_2RuO_4, with arrows standing for the moments on the Ru sites. Upper picture stands for transversal spiral with positive helicity. Combination of two spirals of opposite helicities produces a Spin Density Wave in the LDA ground state, shown on the lower diagram. Some of Ru atoms are circled to emphasize the direction of the density modulation vector $\mathbf{q}_N \parallel [110]$

Calculations for a supercell with three Ru ions reveal an SDW slightly favored over the spiral configuration. Fig. 3 illustrates the distribution of magnetization for both types of spin ordering. An SDW description can be facilitated by introduction of two spin momenta per Ru site, one for a transverse spiral with right helicity as shown on the Fig. 3, and the other with left helicity, assuming this feature is an intrinsic structure of spin density matrix. This approach implies a spin-density matrix consideration on the intra-atomic scale, that is, however, beyond the present numerical implementation. Therefore, an LSDA evaluation of SDW is much more challenging, and we could accomplish the problem only by an appropriate enlargement of the unit cell to fit the nesting vector. As a result, there are two inequivalent Ru sites, one with $M = 0.8\mu_{\mathrm{Bohr}}$ at the equilibrium lattice parameter which is also the amplitude of spiral moment, and the other two just half of that value and opposite in orientation, so that the net magnetization is vanishes. This means that on-site exchange coupling $j_{i,rl}$ between spin moments of spirals with opposite helicites $M_{k,r}$ and $M_{k,l}$, respectively, is essentially smaller than inter-site exchange J_{ij}, given that the following interaction model

$$E^{\mathrm{sdw}} = E^{\mathrm{spiral}} - \sum_{ij}{}' j_{ij,rl} M_{i,r} M_{j,l} - \sum_{i} j_{ii,rl} M_{i,r} M_{i,l}, \tag{4}$$

is relevant. In fact, the last terms in eq. 4 which stand for the exchange interaction between opposite-helicity spirals, are numerically shown to be just 0.2 of the spiral energy. It is also in accordance with the calculations to claim that the on-site interaction between spirals is dominant, since a superposition of only undistorted spirals leads to the formation of the sinusoidal SDW, and therefore the middle term in Eq. 4 can be discarded.

15.3 Formation of spin and orbital moments and pressure dependencies

Now, to better understand the trends under pressure, the mechanism of spin formation has been analyzed in terms of the Anderson impurity model [20] instead of the Stoner approach, more common for itinerant systems. The choice of the framework is motivated by the fact that the formation of a magnetic moment is mainly due to the exchange self-energy arising as the Coulomb repulsion between opposite spins rather than the Stoner-type exchange between parallel spins. It was shown earlier [21] that the electronic structure of pure strontium ruthenate can accurately be reproduced by a tight-binding Hamiltonian which includes three type of bands, since an almost ideal octahedral crystal field

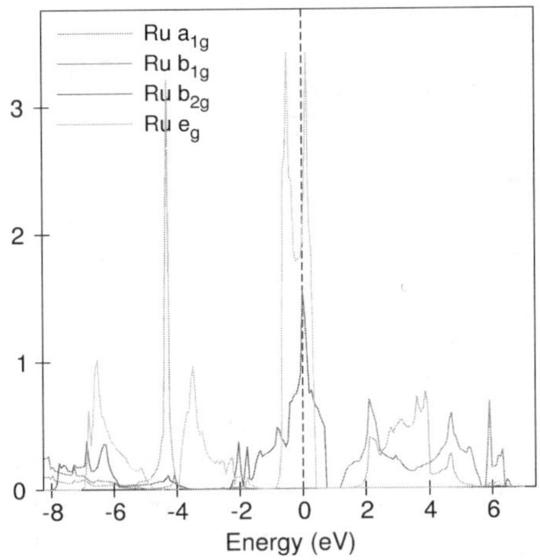

Figure 4. Symmetry-resolved partial density of states on Ru site: tetragonal point symmetry group D_{4h} implies partially (2/3) occupied t_{2g} states split into double-degenerated e_g xz and yz orbitals, and b_{2g} in-plane xy orbital. Shown are also other irreducible states a_{1g} and b_{1g} with z^2 and $x^2 - y^2$ symmetries, clustered however far away from the Fermi level.

splits the d-states of Ru into two groups of states with irreducible representation e_g for two unoccupied and t_{2g} for three partially occupied xz, yz and xy orbitals. With a more accurate treatment, the tetragonal point symmetry group D_{4h}, relevant for tetragonally distorted octahedra (Jahn-Teller distortion), lifts the degeneracy and implies the following irreducible symmetry split of the Ru d-states: as shown on Fig. 4, partially occupied t_{2g} states cast upon double-degenerated e_g-states, with 1.4 occupancy per orbital, and single b_{2g}-orbital, with 1.3 electrons. There are also single orbitals of a_{1g} and b_{1g} symmetry, with 0.95 and 0.8 occupancy respectively, which are, however far from the Fermi level.

The first two degenerate orbitals constitute quasi-one-dimensional bands with a width of about 1 eV and nested around the Fermi level, whereas the third band is two-dimensional and as almost 2 eV broad, with its xy orbital located 0.5 eV below the Fermi level. Therefore, a Hartree-Fock Hamiltonian with two degenerate t_{2g} orbitals xz and yz seems quite relevant for a qualitative description of the moment formation. In this approximation, the condition for moment formation is formulated in terms of four spin occupation numbers $n_{i\sigma}$, two $\sigma = +, -$ per each orbital $i = xz, yz$, Coulomb repulsion between opposite-spin same orbitals U, and "ordinary" exchange interaction between same-spin different orbitals J,

$$\hat{H} = E(\sum_{i\sigma} n_{i\sigma}) + \sum_{\mathbf{k}\sigma} \epsilon_{\mathbf{k}} \hat{n}_{\mathbf{k}\sigma}$$
$$+ \sum_{\mathbf{k}\sigma} v_{i,\mathbf{k}} (\hat{c}^{+}_{i\sigma} \hat{c}_{\mathbf{k}\sigma} + h.c.) + (U - J) \sum_{\sigma} \hat{n}_{xz\sigma} \hat{n}_{yz\sigma}$$
$$+ U \sum_{i} \hat{n}_{i+} \sum_{i} \hat{n}_{i-}. \tag{5}$$

Here i runs over the orbitals, $\sigma = \pm$ is spin projection, E is the unperturbed position of the two t_{2g} levels xz, yz assumed degenerate, $\epsilon_{\mathbf{k}}$ stands for dispersion of itinerant electrons other than the two orbitals, and $v_{i,\mathbf{k}}$ symbolizes the hopping between electrons of the orbitals and the states of the rest. By Hartree-Fock treatment [20], each effective xz, yz level will be spread out into a virtual band of width δ, so that the effective Hartree field, for example, for level $xz+$ becomes

$$E_{xz+} = E + U \sum_{i\sigma}{}' n_{i\sigma} - J n_{yz+}, \tag{6}$$

the prime in the sum means all other than $xz+$, similar for other levels. If the localized states lie close to the Fermi surface, the condition for the formation of spin moment reads

$$(U + J)\eta_{i\sigma}(E_F) \geq 1, \tag{7}$$

where $\eta_{i\sigma}(E_F)$ is the paramagnetic density of states at the Fermi level per spin/orbital equal for all orbitals of any spin. The occupation numbers for $Ru^{4+}$are related to the fact that there are three t_{2g} orbitals equally filled with 4 electrons, i.e. 2/3 per each orbital and spin. There is an obvious relation between the relative Hartree field shift of the orbitals and the band exchange splitting $\bar{\Delta}_{xc}$ averaged over the Brillouin zone, i.e.

$$\bar{\Delta}_{xc} \approx \frac{1}{2} \sum_{i} (E_{i-} - E_{i+}) = \frac{1}{2}(U + J)\delta n, \tag{8}$$

where $\delta n \equiv \sum_i (n_{i+} - n_{i-})$ is the total magnetic moment.

The magnitude of the exchange integral J on Ru site has been estimated to be around 1.0 eV and is practically insensitive to the Coulomb screening effects due to the short-range character of the exchange interaction. On account of $\bar{\Delta}_{xc} \sim 1$ eV, vz. Fig. 5, and $\delta n \sim 0.8$, one obtains for the Coulomb opposite-spin repulsion parameter $U \sim 1.5$ eV at equilibrium lattice parameter, which grows with lattice expansion. In

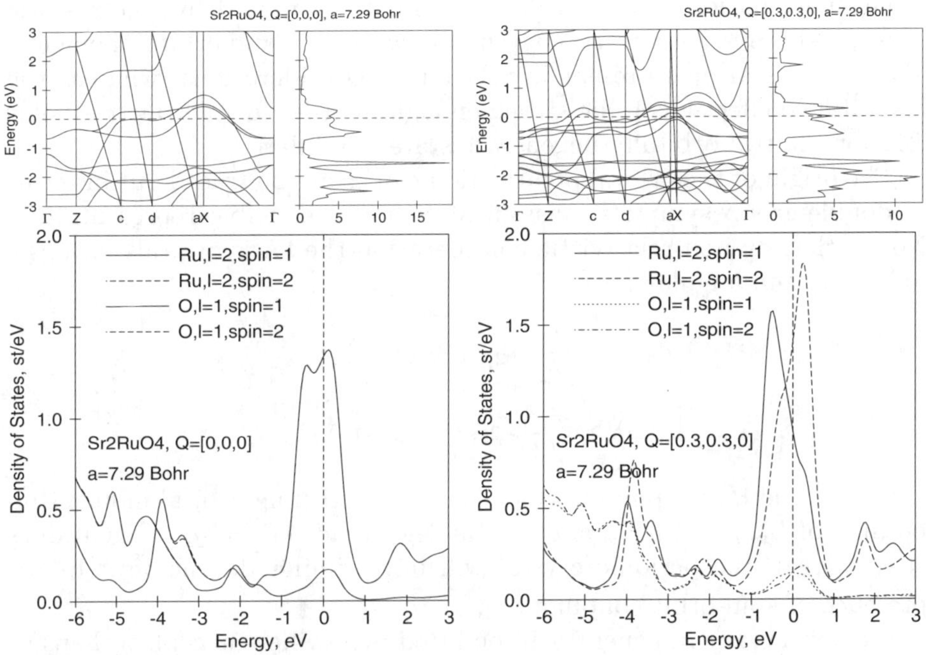

Figure 5. Sr_2RuO_4, electronic structure and projected density of states for spin-restricted and simple spiral configurations, at equilibrium lattice parameter $a = 7.29$ Bohr. Orbitals mostly contributing to the states shaping the nesting of the Fermi surface and causing the local magnetization are mostly of xz and yz angular character, with total occupation of nearly 2.8 electrons per Ru site. The broader states built out of in-plane Ru xy orbitals are occupied by 1.3 electrons.

contrast, U becomes smaller under positive pressure due to an increase in Coulomb screening, which leads to a magnetization collapse with only a 3% reduction of the lattice constant. The spin moment formation condition given by Eq. (7) at the same time reads $(U+J)\eta_{i\sigma}(E_F) \sim 1.65$ since of $\eta_{i\sigma}(E_F) \approx 0.65$ st/eV/spin, which is large enough to satisfy the condition given by Eq. (7) and the exchange splitting if not the effects of quantum fluctuations.

Formation of a local orbital moment requires, along with the spin moment formation, a certain local symmetry imposed by electrostatic crystal fields and relativistic coupling between spin and orbital moments. The latter is always present in relativistic considerations, whereas the crystal-field effect has to be strong enough in order for the Hartree fields E_{i+} for the two orbitally degenerate states to differ.

The origin of the orbital polarization can easily be traced out in terms of double-group symmetry classification for irreducible representation. Notice that by non-relativistic consideration the basis orbitals have the following spin-angular part

$$\mid \phi_{1s}\rangle \sim \frac{i}{\sqrt{2}}(Y_{2,+1}+Y_{2,-1})\hat{\chi}_s,$$
$$\mid \phi_{2s}\rangle \sim \frac{1}{\sqrt{2}}(Y_{2,+1}-Y_{2,-1})\hat{\chi}_s,$$

with $\hat{\chi}$ as the Pauli spinors, which leads to vanishing orbital momentum because of $\hat{L}_z Y_{lm} = mY_{lm}$ and orbital degeneracy $\epsilon_{1s} = \epsilon_{2s'}$. Eigenvalues $\epsilon_{n=1,2s}(\mathbf{k})$ in the periodic crystal will have similar degeneracy. In the presence of spin-orbit coupling $\hat{h}_{so}$,

the degeneracy in general will be lifted everywhere except perhaps at some high-symmetry points in the Brillouin zone. By construction of single-particle eigenvectors from Bloch combinations of the orbitals, the respective eigenvalues can in lowest perturbation order, easily be shown to be split at every **k**-point

$$\bar{\epsilon}_{ns}(\mathbf{k}) = \epsilon_{(+)} + \mathbf{s}\sqrt{\epsilon^{\mathbf{2}}_{(-)} + \xi^{\mathbf{2}}}, \tag{9}$$

where the spin-orbit coupling parameter $\xi^2 = \mid \langle \psi_{1s} l_z \psi_{2s} \mid\rangle^2 + \mid \langle \psi_{1+} l_+ \psi_{2-}\rangle \mid^2$, and $\epsilon_{(+)}$, $\epsilon_{(-)}$ stand for half-sum and half-difference of the non-relativistic bands ϵ_{ns}, respectively. Thus, electrostatic crystal fields and the spin-orbit coupling split the energy levels

$$\bar{\epsilon}_{ns} \approx \epsilon + s \mid \xi \mid \tag{10}$$

at general k-point, with $s = \pm$ (not a spin index anymore) and $\xi = \langle \phi_{1s} \hat{l}_z \phi_{2s'} \rangle$ since $\langle \hat{l}_{\pm} \rangle$ is vanishing on the Ru sites. The corresponding eigenvectors in the lowest order of perturbation theory are

$$| \mathbf{k}ns\rangle = \sum_{is'} \alpha_{ns}^{is'}(\mathbf{k}) \mid \phi_{is'}\rangle, \tag{11}$$

which for the angular part implies

$$| \mathbf{k}1s\rangle \sim Y_{2,+1}\hat{\chi}_s,$$
$$| \mathbf{k}2s\rangle \sim Y_{2,-1}\hat{\chi}_s,$$

since α_{is} is either one or zero if no magnetic field or exchange splitting is present. More formally this result could be obtained within double-group classification when the wave function is to be expanded in terms of 4 basic functions corresponding to Δ_6^5 and Δ_7^5 irreducible representations of C_{4v} group [22].

By expanding the field operators in terms of the *is*-orbitals $\hat{\psi} = \sum_{ns,is'} \alpha_{ns}^{is'}(\mathbf{k}) \mid \mathbf{k}is'\rangle \hat{c}_{ns}(\mathbf{k})$, one easily verifies that the projection of the net orbital momentum $\langle \hat{\mathbf{L}} \rangle_\zeta \equiv \hat{\mathbf{e}}_\zeta \langle \hat{\psi} \mid \mathbf{l} \mid \hat{\psi} \rangle$ along spin quantization direction $\hat{\mathbf{e}}_\zeta$ will be different from zero only if the spin degeneracy of each state is lifted. Then, in the lowest order of the perturbation theory in spin-orbit coupling [23–25]

$$\langle \hat{\mathbf{L}} \rangle_\zeta \approx -4\xi \sum_{1234,\sigma} A(12,34,\sigma\sigma) \langle 1\sigma \mid \hat{\mathbf{e}}_\zeta \mathbf{l} \mid 2\sigma \rangle \langle 3\sigma \mid \sigma \mathbf{l} \mid 4\sigma \rangle$$
$$\equiv \hat{\mathbf{e}}_\zeta (\langle \mathbf{L} \rangle^+ + \langle \mathbf{L} \rangle^-), \tag{12}$$

where the matrix element

$$A(12,34,\sigma\sigma') \equiv \int_{\epsilon<E_F<\epsilon'} \frac{d\epsilon d\epsilon'}{\epsilon-\epsilon'} \sum_{\mathbf{k}} n_{12,\sigma}(\mathbf{k},\epsilon) n_{34,\sigma'}(\mathbf{k},\epsilon')$$

is determined through the density of states $n_{ii',\sigma} = \sum_n a_{i\sigma *} a_{i'\sigma} \delta(\epsilon - \epsilon_{n\sigma}(\mathbf{k}))$ with non-relativistic eigenenergies $\epsilon_{n\sigma}(\mathbf{k})$ and expansion coefficients $a_{i'\sigma}$, and remains unchanged for any orientation of the spin quantization axis. The orbital moment $\langle \hat{\mathbf{L}} \rangle_\zeta$ depends on the difference of the Hartree field for the *same*-spin states (crystal-field splitting), the amount of spin-orbit splitting $\mid \xi \mid$, which is about 0.15 eV for the Ru $4d$-orbitals, as well as the orbital-projected spin and particle DOS. At the same time, the magneto-crystalline energy

$$\delta E \approx -\frac{1}{4} \xi \hat{\mathbf{e}}_\zeta (\langle \mathbf{L} \rangle^+ - \langle \mathbf{L} \rangle^-) + (\mathrm{spin} \ - \ \mathrm{flip\,terms}), \tag{13}$$

is related to the difference between same orbital moment spin-components, given that the spin-flip effects are small. The angular dependence is given by $\langle i\sigma \mid \mathbf{l}\sigma \mid i'\sigma \rangle$ only, which in case of uniaxial symmetry is simply $K_0 + K_1 \sin^2 \theta$. This specifically means that the same

Table 1. SDW configuration: magnetization, spin S_ς and orbital L_ς contributions at different orientation of the spin quantization axis $\hat{\mathbf{e}}_\varsigma$, in units of μ_{Bohr}. O_{ei} and O_{ai} refer to oxygens equatorial and apical in respect to Ru1 and Ru2, respectively.

Atom	$\hat{\mathbf{e}}_\varsigma$||[001]		$\hat{\mathbf{e}}_\varsigma$||[010]		$\hat{\mathbf{e}}_\varsigma$||[110]	
	S_ς	L_ς	S_ς	L_ς	S_ς	L_ς
Ru1	+0.832	+0.007	+0.746	+0.013	+0.748	+0.019
Ru2	-0.393	-0.010	-0.393	-0.002	-0.400	-0.000
O_{e1}	+0.032	+0.002	+0.025	+0.001	+0.024	+0.002
O_{e2}	-0.053	-0.003	-0.055	-0.003	-0.055	-0.002
O_{a1}	+0.042	+0.005	+0.042	+0.004	+0.042	+0.003
O_{a2}	-0.023	-0.003	-0.022	-0.002	-0.022	-0.002

matrix elements are present at *any* orientation of spin quantization axis. From the formulas given by Eqs. 12 and 13, it is easy to project a similar scaling of orbital momentum and energy if the shell is more than half-filled [24], since the orbital polarization of spin-majority states has to vanish. Our LSDA-DFT calculations find an opposite orientational trend, which is obviously related to the effects of hybridization between *opposite*-spin states. Notwithstanding a remarkable spin anisotropy with data collected in the Table 1, the energy shows practically *no* dependence on the orientation of the quantization axis. The in and out of plane difference is less than 0.01mRy, whereas the spin moment shows considerable out of plane enhancement. We will use this fact to illustrate the destructive role of quantum spin fluctuations on long-range ordering. Note also that it follows from Eq. 12, considering only two orbitals would give $K_0 = K_1$. This would imply a factor of two difference polarization of a factor of two between in-plane and out-of-plane orientation of the spin quantization axis, which only qualitatively correlates with the numbers from the Table 1, which includes hybridization effects between *all* orbitals. Therefore, there is only limited legitimacy for the two-orbital model by interpretation of the orbital moment formation.

The sign of the projection is obviously a matter of the local occupation (3-rd Hund's rule) and in accord with the Eq. (12) the exchange-caused $\langle \hat{l}_z \rangle$ will be pointed in the direction of spin moment, which is common for orbitals with greater than half-filling. On the other hand, for the numbers for orbital occupation and spin magnetization caused by an *external* magnetic field, $\bar{l}_z$ is pointed against the spin vector, since *all* orbitals will now be involved in the magnetic response, with an effective occupation *less* than half-filling for $Ru^{4+\cdot}$ ion.

Variation of the volume causes a certain trend in the amplitude and stability of the spiral configuration. As illustrated in Fig. 6, there is

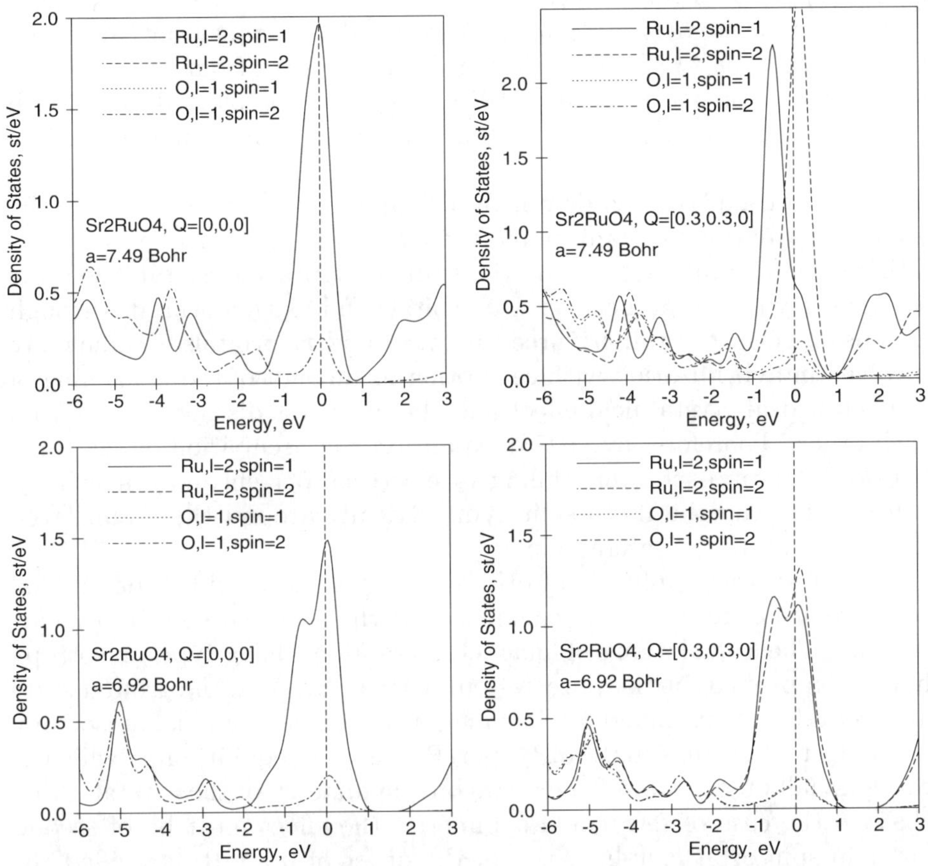

Figure 6. Sr_2RuO_4, projected density of states for spin-restricted and simple spiral configurations, at extended and reduced lattice parameters $a = 7.49$ and $a = 6.92$ Bohr, respectively. Local magnetization develops obviously in accord with the amplitude of *pd* overlap between cations and anions, what particularly leads to a collapse of magnetism at the lattice contraction.

obviously a higher paramagnetic density of states $\eta_\sigma(E_F) = 1.8$ st/eV with lattice expansion of 3%, which in accord with the Eq. (7) facilitates spin formation with an increase of the moment $M = 1.0\mu_{\rm Bohr}$. With lattice compression of the same amount, the increased hopping between Ru and O ions leads to a diminishment of the DOS, and the condition of Eq. (7) is no longer fulfilled.

That the system is close to a metal-insulator transition caused by the formation of AFM spin ordering, is clearly seen from Fig. 7, where the electronic structure of an SDW configuration is plotted along with the spin- and atom-projected DOS at equilibrium and expanded lattice parameter.

The AFM gap itself is driven by the aforementioned local magnetic moment formation mechanism and the long-range nesting-imposed oscillating exchange interaction. As the spin configuration is stabilized due to covalency reduction between Ru orbitals and oxygen orbitals through expansion of the crystal, *all* three Ru xz, yz and xy orbitals become more involved in hybridization with each other as a result of the weakening of the octahedral crystal field effect and therefore less distinguishible from each other. Therefore, by lattice expansion the occupation of the three Ru orbitals approaches half-filling (4 electrons per entire t_{2g}-subshell), causing the Ru orbitals to split symmetrically around the Fermi level, due to the exchange interaction.

Of course, the amplitude of orbital polarization is still to be verified by more accurate numerical calculations, which indeed confirm the situation. The calculations of magneto-crystalline Ising-type anisotropy have been probed for the pure system with an SDW in the ground state at several lattice parameters. The magnitude of the uni-axial anisotropy amounts to less than 0.01 mRy per Ru unit at equilibrium, which is rather small on account of the tetragonal symmetry of the crystal. Note that, in the case of tetragonal symmetry, the effect must be of second order in spin-orbit coupling, i.e. of the order of 0.1 mRy provided the second Hund's rule is fulfilled. But this is not the case, as we already deduced on the basis of the impurity model. The situation changes only slightly with variation of the lattice parameter around the equilibrium: while remaining in the area of tetragonal elongation of the octahedra, the spin alignment starts to stabilize by increased anisotropy energy only with considerable lattice expansion of 3%. This behavior shows that the quenching of the orbital moment is a characteristic feature of ideal strontium ruthenate within a range of tetragonal-symmetry-conserving lattice distortion around the equilibrium. Upon octahedral rotation, orthorhombic crystal fields will likely enhance the spin and orbital polarization. Although considered speculation, this dependence of orbital

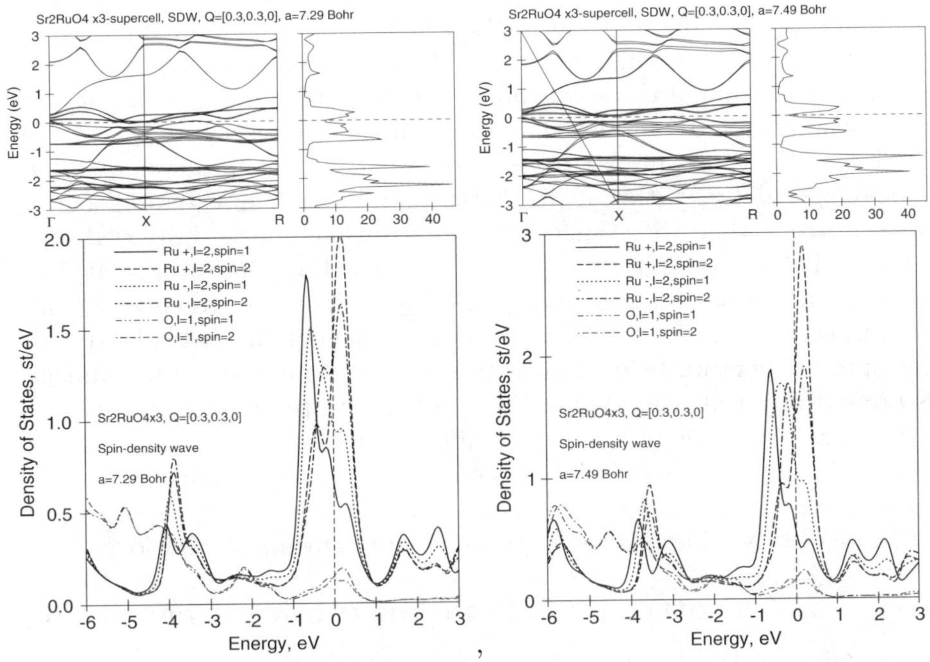

Figure 7. Sr_2RuO_4, electronic structure and projected density of states for spin-density wave configuration (ground state), for equilibrium lattice with $a = 7.29$ Bohr and extended lattice with $a = 7.49$ Bohr. The most prominent feature is the opening of the gap on the density of states at the Fermi level, what gives evidence for an AFM-caused metal-insulator trend in the pure compound.

polarization on tetragonal strain can at least partially be used to explain why the moments induced by Ti^{4+} on the Ru sites exhibit strong Ising anisotropy in the experiment [28]. Of course, by partial replacement of Ru by Ti the octahedral distortion is actually a local effect and can't properly be simulated by a simple uniform lattice expansion or compression as used in our attempt to describe the respective trend in the formation of the orbital moment.

The vanishing value for the anisotropy at equilibrium is certainly crucial for the formation of the magnetic moment since zero-point spin fluctuations will now destabilize the ground state SDW in the absence of a mechanism for pinning the orientation of local moments. This effect of renormalization of the local spin momentum by quantum fluctuations was considered in detail for 2D antiferromagnetic Heisenberg systems on the basis of a $1/S$ spin expansion, see e.g. [26, 27], and it was asserted, though speculatively, that if the "trial" spin is small enough and there is no anisotropy then the staggered magnetization can be destroyed completely. There are rigorous results [29] which show that at $T = 0$ there is long-range order for spin moments $S \geq 1$. There is growing evidence that in case of the 2D Heisenberg antiferromagnet the critical value for the spin momentum below which the staggered magnetization vanishes is around physically accessible $S_c \sim 1/2$ within the model. In our case with an SDW in the ground state, and the hydrodynamic description as a more appropriate model, this value shoud be even less. The largest spin moment on the Ru site 0.4 is less than the Heisenberg critical value and most likely will collapse in presence of quantum fluctuations.

15.4 Anisotropy of static magnetic susceptibility

We also checked the magnitude of the paramagnetic static spin susceptibility, which is related to the DOS at the Fermi level, and found reasonable agreement with the available experimental data, obtained from neutron scattering. For an applied magnetic field of $H = 1$ T, the calculated induced total spin moment per unit cell turns out to be $M_{ind} = 1.15 \times 10^{-3}\mu_{Bohr}$, which corresponds to $\chi_{isotr}(\mathbf{q} = 0, \omega = 0) \approx 1.15 \times 10^{-3}[\mu_{Bohr}/T] = 6.64 \times 10^{-4}$emu/mol for the isotropic susceptibility. This theoretical value for the non-relativistic and therefore spatially isotropic χ correlates well with experimental values which are in the range of $5.29 \times 10^{-4} - 9.0 \times 10^{-4}$ emu/mol [12, 13]. Given that the susceptibility is proportional to the DOS at the Fermi level, this seems to have a reasonable magnitude. Therefore, the observed effect of local spin polarization is obviously imposed not by the high value of the DOS, which otherwise would be an artifact had the DOS been overestimated in

Table 2. Static paramagnetic susceptibility $\chi(\omega = 0, \mathbf{q} = 0)$ in units of $10^{-4}\mu_{\text{Bohr}}/\text{T}$, determined as a response to an external magnetic field applied along [001], [010], and [110] axis. The shown are partial spin χ_S and orbital χ_L contributions from Ru and oxygens. $O_{e1,e2}$ and O_{ap} stand for oxygens equatorial and apical in respect to Ru, respectively. For symmetry reason, the equatorial oxygens are different only at $\mathbf{H} \parallel [010]$.

Atom	$\mathbf{H}\|\|[001]$		$\mathbf{H}\|\|[010]$		$\mathbf{H}\|\|[110]$	
	χ_S	χ_L	χ_S	χ_L	χ_S	χ_L
Ru	+8.3	-0.2	+10.0	-0.9	+10.0.	-0.9
O_{e1}	+1.3	+0.1	+1.6	+0.1	+1.7	-0.2
O_{e2}	+1.3	+0.1	+1.7	-0.4	+1.7	-0.2
O_{ap}	+0.2	0.0	+0.2	0.0	+0.2	0.0
Sr	+0.2	0.0	+0.2	0.0	+0.2	0.0

calculations, but rather by the Ru on-site exchange enhancement, which is responsible for the first Hund's rule.

Inclusion of spin-orbit coupling causes a double anisotropy effect. First, there is a very modest magneto-crystalline anisotropy of about less than 0.01 mRy as mentioned above. Second, the static paramagnetic longitudinal susceptibility also gains a small anisotropy $\chi_{ab}(\mathbf{q} \sim 0)/\chi_c(\mathbf{q} \sim 0) \sim 1.09$ at vanishing $\mathbf{q}$ vectors. This means the amplitude of spin fluctuations is slightly enhanced within the plane. The partial spin and orbital contributions, to the total magnetic susceptibility, resolved also in consituents of octahedron, is given in Table 2 (numbers are rounded). For symmetry reasons, equatorial oxygens along [100]- and [010]-bonds with Ru labeled as O_{e1} and O_{e2} respectively, are not equivalent at $\mathbf{H} \parallel [010]$.

Note from Table 2 that there is considerable orbital polarization of the equatorial oxygen ions in the presence of an external field, which makes the Ru ions orbitally polarized against the local spin moment, no matter what the orientation of the field. On the other hand, apical oxygens and Sr have very low magnetic polarizability, which again justifies the 2D-character of the Fermi system and particularly makes the inter-planar exchange coupling immaterial. This fact also indicates that incorporation of orbitals on both cation and anion sites in any analytical model of spin susceptibility is mandatory in the perovskite. This could explain the unrealistically large anisotropy [14] of the magnetic susceptibility obtained within an RPA-designed two-band Hubbard model similar to the one used above. The two-band model seems, however, quite adequate for a qualitative discussion of the orientational effect, since in the case of spontaneous polarization, oxygens are practically uninvolved in the low-energy angular momentum dynamics.

To reflect the aforementioned effects of anisotropy, the following formula

$$\mathbf{M} = \chi_{\|}\mathbf{H}_{\|} + [\chi_{\perp} + \kappa(h_{\perp,x}h_{\perp,y})^2]\mathbf{H}_{\perp} \tag{14}$$

can be used to adequately describe the basal-plane anisotropy of magnetic response with local C_{4v} symmetry and uniaxial anisotropy along c-axis. The notations used are: $\|$, $\perp$ stand for components parallel/perpendicular to c-axis, e.g. external field $\mathbf{H} = \mathbf{H}_{\|} + \mathbf{H}_{\perp}$, and in-plane unit-vector $\mathbf{h}_{\perp} \equiv \mathbf{H}_{\perp}/H_{\perp}$. Thus

$$\begin{aligned} \mathbf{M} &= \chi_{\|}\mathbf{H}, \qquad \mathbf{H} \parallel c, \\ \mathbf{M} &= \chi_{\perp}\mathbf{H}, \qquad \mathbf{H} \parallel [100], \\ \mathbf{M} &= (\chi_{\perp} + \frac{1}{4}\kappa)\mathbf{H}, \qquad \mathbf{H} \parallel [110], \end{aligned} \tag{15}$$

the latter set is the most favorite for polarization, apparently because of the spin-orbit interaction causing most of the charge transfer between cations and anions when the field is along [110]. The induced orbital and spin moments on Ru sites are antiparallel, an expected result in view of the fact that the entire Ru^{+4} d-shell is less than half-filled. This observation is in contrast to the prediction of the two-orbital model discussed above in accord with which the orbital L and spin S moments would be expected to be collinear since the 2/3 occupation of the orbitals is clearly above the half. In fact, L and S are collinear by spontaneous spin polarization of Ru ions in the incommensurate SDW phase, which shows that the on-site exchange field responsible for the respective spontaneous splitting is contained mainly within the two considered Ru orbitals as having the highest Coulomb repulsion U between the same spins. When a uniform external field is applied, the Zeeman interaction acts on *all* sites and *all* orbitals, thus involving the *entire* d-shell on the Ru^{+4} sites with effective occupation of about 4 electrons (less than half-filling), as mentioned above. In this case the orbital polarization has to adopt the minimum total moment $J = | S - L |$ within the manifold of the Lande levels of the given L, S multiplet. By selectively switching on spin-orbit coupling parameter on various ions it is established that the dominant relativistic effect comes from the d-orbitals of Ru, also the polarizing the neighbour equatorial oxygens. The polarizability on the sites of apical oxygens and adjacent ions of Sr^{+2} is vanishingly small, presumably as a result of complete 'undressing' of the Sr ions.

The following ratios for the susceptibility components

$$\chi_{\perp} \approx 1.09\chi_{\|},$$

$$\frac{1}{4}\kappa \approx 0.01\chi_{\|}, \tag{16}$$

complete the anisotropy mapping of the magnetic susceptibility at zero wave vector. So far we have discussed the anisotropy found for static χ at zero wave vector, particularly emphasizing stronger ferromagnetic fluctuations within the *ab*-plane. Since the anisotropy of χ is closely related to anisotropy of $\langle l_z \rangle_\zeta$ the same magnitude of anisotropy would be expected for finite $\mathbf{q}$ as well. From the above discussion of the orientational dependence in Sec. 15.4, it is clear that the effect of the energy denominator which *doesn't* depend on the orientation of spin quantization axis (external magnetic field), will affect the susceptibility at the nesting vector leading to an enhancement of both longitudinal χ_{zz} and transversal χ_{+-} components in rather *similar* way. This conclusion is in contradiction to the earlier prediction [14] that χ_{+-} has no features related to the Fermi surface nesting whereas χ_{zz} does. Apparently only matrix elements of $\hat{l}_z$ are contributing to magnetic susceptibility within the two-orbital model, and they do that at *any* orientation of $\hat{\mathbf{e}}_\zeta$. For wave vectors around the nesting $\mathbf{q}_N$, both paramagnetic longitudinal and transverse components are expected to be enhanced, with a rather modest level of anisotropy, since this is solely an effect of the energy denominator.

15.5 Summary: possible magnetic low-lying excitations and impact upon superconductivity

The presence of a SDW ground state, although suppressed by zero-point fluctuations as mentioned above, raises questions about specific magnetic low-lying excitations, a somewhat general concern that might have some interesting consequences for the observables. Establishment of magnetic type low-lying excitations becomes even more intriguing with reference to potential mediators of superconducting pairing since earlier lattice dynamics investigation [4] shows no sign of phonon mode softening.

We have demonstrated that there is a certain trend toward the formation of a spin-density wave already in pure stronthium ruthenate. On the other hand, in view of the immaterial magneto-crystalline anisotropy the quantum spin fluctuations are most likely to destroy the staggered magnetization density. Yet, the magnetic susceptibility is expected to be structured by a type of low-lying magnetic excitations, which in spin-wave approximation, can be considered as phasons with limited life-time. The latter type of excitations is characteristic of a simple helix in the ground state. The problem was first considered some time ago [17, 18] within a Heisenberg model and recently within LSDA [19] in conjunction with the heavy rare earth metals, where the ground and meta-stable states are vary from ordinary FM through the AFM helix and FM cone. The dispersion of phasons, like antiferromagnetic magnons, is linear in wave vector and has no gap, since in presence of incommensurate spin modulation the azimuthal phase is averaged out from the total energy. The phasons exemplify the Goldstone mode as a result of spontaneous symmetry breaking in the system.

There is the question of whether the phasons, perhaps better termed para-phasons, might be involved in the superconductivity. If they are, then the classification in terms of singlet or triplet would be unnecessary, since on atomic scale the spinor symmetry is triplet, but on the scale of the helix the symmetry is a singlet. Although, as argued above, a vanishingly small magnetocrystalline anisotropy on the Ru sites cannot stabilize the long range order in the presence of zero-point spin motion, the latter itself has to reflect the symmetry of incommensurate helical superstructure or spin-density wave.

Since the observation of a temperature-independent spin susceptibility deep into the superconducting state has been used to justify a time-reversal symmetry breaking triplet order parameter within Cooper pairing formalism, a detailed analysis shows that the "chiral" vector order developed for ^{3}He is also relevant for the crystal in question (see Ref. [30] for a thorough discussion of this and related problems). Here we restrict ourselves to a brief discussion of possible complications of our results for the triplet pairing models.

Similar to the B phase of ^{3}He, broken symmetry in the unconventional superconductor Sr_2RuO_4 governs the relative orientation of spin and orbital momentum. The on-site exchange on Ru sites seems to be strong enough to produce local ferromagnetic fluctuations of spin density, correlated in helical or spin-density wave structure by the RKKY-type interaction, which is a clear manifestation of the crystalline structure. On the other hand, the relativistic interaction is large enough to keep spin and orbital moments together up to 1000 K, but the vanishingly small anisotropy decouples the entire set of spin-orbit pairs from certain directions in the crystal. Therefore the superconducting order parameter seems similar to the one in the B phase of ^{3}He, but modulated spatially as a helical or spin-density wave order parameter. Here we want to emphasize once again, that the magnitude of spin-orbit coupling alone cannot determine the anisotropy of the order parameter. Instead, the SO effect has to be combined with the subtle effects of crystal field and state occupation, which in our particular case leads to the vanishingly small anisotropy of magnetic fluctuations. Note also, as a support for our finding, that there is so far no experimental proof that the vector order parameter is pinned along a certain direction in the crystal.

Clearly, this observation based on the exchange structure of the spin interaction, has to be complemented by the anisotropy mapping of the magnetic susceptibility which is particularly suited to the liquid ^{3}He B phase model of superfluidity for the tetragonal crystalline structure of strontium ruthenate. By giving a credit to LSDA calculations, we have to admit that the zero wave-vector spin fluctuations are at maximum within the plane of the essentially 2D Fermi system along the [110] Ru-chain resembling the A phase of ^{3}He, and strong enhancement of fluctuations at the Fermi surface nesting vector with rotationally broken symmetry as in the ^{3}He B phase due to the vanishingly small orientational anisotropy in that case.

Acknowledgments

The authors are grateful to M. Johannes, A. Shabaev, I. Mazin, M. Pederson and B. Gyorffy for useful discussions.

References

[1] F. Servant *et al*, Phys. Rev. B **65**, 184511 (2002).

[2] M. Braden, W. Reichardt, S. Nishizaki, Y. Mori, and Y. Maeno, Phys. Rev. B **57**, 1236 (1998).

[3] O. Chmaissem, J. D. Jorgensen, H. Shaked, S. Ikeda, and Y. Maeno, Phys. Rev. B **57**, 5067 (1998).

[4] S. V. Halilov and D. J. Singh, Phys. Rev. B (2002).

[5] R. Matzdorf *et al*, Science **289**, 746 (2000).

[6] K.M. Shen *et al*, Phys. Rev. B **64**, 180502(R) (2001)

[7] M. Braden, O. Friedt, Y. Sidis, P. Bourges, M. Minakata, and Y. Maeno, Phys. Rev. Lett. **88**, 197002 (2002).

[8] T. Oguchi, Phys. Rev. B **51**, 1385 (1995)

[9] D. J. Singh, Phys. Rev. B **52**, 1358 (1995).

[10] D. H. Lyons and T. H. Kaplan, Phys. Rev. **120**, 1580 (1960)

[11] P. W. Anderson, in *Solid State Physics*, ed. by F. Seitz and D. Turnbull, v. 14, Academic Press, NY 1963

[12] J. J. Neumeier *et al*, Phys. Rev. B **50**, 17910 (1994).

[13] Y. Maeno *et al*, J. Phys. Soc. Jpn. 66, 1405 (1997).

[14] I. Eremin, D. Manske, and K. H. Bennemann, Phys. Rev. B, **65**, 220502 (2002).

[15] H. Mukuda et al., J. Phys. Soc. Jpn. **67** 3945 (1998).

[16] S. V. Halilov, H. Eschrig, A. Perlov, and P. M. Oppeneer, Phys. Rev. B **58** 293 (1998).

[17] T. A. Kaplan, Phys. Rev. **109**, 782 (1957).

[18] A. Yoshimori, J. Phys. Soc. Jpn. **14**, 807 (1959).

[19] A. Y. Perlov, S. V. Halilov, and H. Eschrig, Phys. Rev. B **61**, 4070 (2000).

[20] P. W. Anderson, Phys. Rev. **124**, 41 (1961).

[21] I. I. Mazin and D. J. Singh, Phys. Rev. Lett. **79**, 733 (1997); A.Liebsch and A. Lichtenstein, Phys. Rev. Lett. **84**, 1591 (2000); D.K. Morr, P. F. Trautman, and M. J. Graf, Phys. Rev. Lett. **86**, 5978 (2001).

[22] J. F. Cornwell, *Group Theory and Electronic Energy Bands in Solids*, North-Holland, Amsterdam, 1969.

[23] H. Takayama, K. P. Bohnen, and P. Fulde, Phys. Rev. B **14**, 2287 (1976).

[24] P. Bruno, Phys. Rev. B **39**, 865 (1989).

[25] G. van der Laan, J.Phys.: Cond. Matt. **10**, 3239 (1998).

[26] P. W. Anderson, Phys. Rev. **86**, 694 (1952); Mater. Res. Bull. **8**, 153 (1973).

[27] S. Chakravarty, B.I. Halperin, and D. R. Nelson, Phys. Rev. B **39**, 2344 (1989).

[28] M. Minakata and Y. Maeno, Phys. Rev. B **63**, 180504R (2001).

[29] E. J. Neves and J. F. Peres, Phys. Lett. **114A**, 331 (1986); F. J. Dyson, E. H. Lieb, and B. Simon, J. Stat. Phys. **18**, 335 (1978).

[30] A.P. Mackenzie and Y. Maeno, Rev. Mod. Phys. **75** 657 (2003).

Index